TABLE OF CONTENTS

Electromagnetic Propagation in Mixed Media
  N.W. ASHCROFT 1

Resonances in the Bulk Properties of Composite Media –
Theory and Applications
  D.J. BERGMAN 10

Elastic Waves in Fluid-Saturated Porous Media
  J.G. BERRYMAN 38

Biot's Poroelasticity Equations by Homogenization
  R. BURRIDGE, J.B. KELLER 51

Approximations of Brinkman Type
  S. CHILDRESS 58

Topology, Geometry, and Physical Properties of Porous Rocks
  M.H. COHEN, C. LIN 74

Wave Propagation in Bubbly Liquids
  D.A. DREW, LAP-YAN CHENG 85

Elastodynamics of Porous Media
  D.L. JOHNSON 97

Bounds for the Effective Conductivity of Random Media
  W. KOHLER, G.C. PAPANICOLAOU 111

Structural Design Optimization, Homogenization and
Relaxation of Variational Problems
  R.V. KOHN, G. STRANG 131

Coherent Medium Approach to Hopping Conduction
  M. LAX, T. ODAGAKI 148

Nonlinear Evolution Equations with Rapidly
Oscillating Initial Data
  D. McLAUGHLIN, G. PAPANICOLAOU, O. PIRONNEAU 177

A Comparison of Two Methods for Deriving Bounds on
the Effective Conductivity of Composites
  G.W. MILTON, R.C. McPHEDRAN 183

Fluctuation Corrections to the Mean Field
Description of a Nonuniform Fluid
  J.K. PERCUS 194

Fingering in Porous Media
  P.G. SAFFMAN 208

On the Effective Thermal Conductivity and
Permeability of Regular Arrays of Spheres
  A.S. SANGANI, A. ACRIVOS 216

Dielectric and Acoustic Response of Rocks
  P.N. SEN 226

Effective Dielectric Function of Composite Media
  PING SHENG 239

Macroscopic and Microscopic Fields in Electron and Atom Transport
    R.S. SORBELLO 251

Propagation and Attenuation in Composite Media
    V. TWERSKY 258

Frequency Dependent Dielectric Constants of
Discrete Random Media
    V.V. VARADAN, V.N. BRINGI, V.K. VARADAN 272

A Variational Method to Find Effective Coefficients
for Periodic Media. A Comparison with Standard Homogenization
    M. VOGELIUS 285

Effective Medium Approximation for Diffusion
on Random Networks
    I. WEBMAN 297

List of Participants 305

Lecture Notes in Physics

Edited by J. Ehlers, München, K. Hepp, Zürich
R. Kippenhahn, München, H. A. Weidenmüller, Heidelberg
and J. Zittartz, Köln

154

Macroscopic Properties of Disordered Media

Proceedings of a Conference
Held at the Courant Institute
June 1–3, 1981

Edited by R. Burridge, S. Childress, and G. Papanicolaou

Springer-Verlag
Berlin Heidelberg New York 1982

Editors

R. Burridge
S. Childress
G. Papanicolaou
Courant Institute, New York University
New York, NY 10012, USA

ISBN 3-540-11202-2 Springer-Verlag Berlin Heidelberg New York
ISBN 0-387-11202-2 Springer-Verlag New York Heidelberg Berlin

This work is subject to copyright. All rights are reserved, whether the whole or part of the material is concerned, specifically those of translation, reprinting, re-use of illustrations, broadcasting, reproduction by photocopying machine or similar means, and storage in data banks. Under § 54 of the German Copyright Law where copies are made for other than private use, a fee is payable to "Verwertungsgesellschaft Wort", Munich.

© by Springer-Verlag Berlin Heidelberg 1982
Printed in Germany

Printing and binding: Beltz Offsetdruck, Hemsbach/Bergstr.
2153/3140-543210

CONFERENCE ON THE MACROSCOPIC PROPERTIES OF DISORDERED MEDIA

June 1-2-3, 1981

COURANT INSTITUTE, NEW YORK UNIVERSITY, 251 MERCER STREET,
NEW YORK, N.Y. 10012

MONDAY JUNE 1

Time	Speaker
9:00 - 9:40	R. Burridge, Courant Institute, Poroelastic equations from microstructure
9:40 - 10:20	P. Saffman, Cal. Inst. Tech., Fingering in porous media
10:20 - 10:40	BREAK
10:40 - 11:20	J. Berryman, Bell Laboratories, Elastic waves in fluid-saturated porous media
11:20 - 12:00	S. Childress, Courant Institute, Approximations of Brinkman type
12:00 - 1:30	LUNCH
1:30 - 2:10	R. O'Connell, Harvard Univ., Porous media, self-consistent methods
2:10 - 2:50	D. Johnson, Schlumberger-Doll, Elastodynamics in porous fluid-saturated solids
2:50 - 3:10	BREAK
3:10 - 3:50	V. Twersky, Univ. of Illinois, Chicago Circle, Propagation and attenuation in composite media
3:50 - 4:30	J. Lebowitz, Rutgers Univ., Kinetics of cluster growth in quenched alloys
4:30 - 5:10	L. Tartar, Univ. of Paris Orsay, Optimal bounds in homogenization and applications
5:10 - 5:50	D. J. Bergman, Tel Aviv University, Resonances in the bulk properties of composite media -- theory and applications

TUESDAY JUNE 2

Time	Speaker
9:00 - 9:40	N. Ashcroft, Cornell University, Electromagnetic propagation in mixed media
9:40 - 10:20	Ping Sheng, Exxon Res., 'Structural units' and the effective medium calculation of composite dielectric constants
10:20 - 10:40	BREAK
10:40 - 11:20	G. Papanicolaou, Courant Institute, Upper and lower bounds for conductivities in random media
11:20 - 12:00	I. Webman, Courant Institute, Diffusion on random networks
12:00 - 1:30	LUNCH
1:30 - 2:10	M. Cohen, University of Chicago, Topology, geometry and physical properties of porous rocks
2:10 - 2:50	M. Lax, City University of N.Y., Coherent medium approach to hopping conductivity

CONFERENCE PROGRAM, continued

Tuesday, continued

2:50 - 3:10	BREAK
3:10 - 3:50	P. Sen, Schlumberger Res., A self-similar model for acoustic and electrical response in rods
3:50 - 4:30	T. Odagaki, City University of N.Y., Hopping conduction in one dimensional chains
4:30 - 5:10	M. Vogelius, Courant Institute, A projection method applied to diffusion in a periodic structure

WEDNESDAY JUNE 3

9:00 - 9:40	A. Acrivos, Stanford University, Slow flow and heat transfer past isotropic arrays of spheres and cylinders
9:40 - 10:20	J. Percus, Courant Institute, Fluctuation corrections to the mean field description of a nonuniform fluid
10:20 - 10:40	BREAK
10:40 - 11:20	J. B. Keller, Stanford University, Reflection and scattering from rough surfaces
11:20 - 12:00	R. V. Kohn, Courant Institute, Homogenization as a tool for engineering design optimization
12:00 - 1:30	LUNCH
1:30 - 2:10	D. Drew, Rensselaer Poly. Inst., Wave propagation in a bubbly liquid
2:10 - 2:50	W. Kohler, Virginia Poly. Inst. and State Univ., Upper and lower bounds for effective parameters
2:50 - 3:10	BREAK
3:10 - 3:50	V.V. and V.K. Varadan, Ohio State Univ., Computation of the frequency dependence of the average dielectric constant in discrete random media
3:50 - 4:30	D. McLaughlin, Courant Inst. and Univ. of Arizona, Self consistent convection of flows with microstructure
4:30 - 5:10	G. W. Milton, Cornell University, A comparison of two methods for deriving bounds on the effective conductivity of composites

The conference is sponsored by a grant from the Exxon Research Corp.

EDITORS' PREFACE

During the first three days of June, 1981, a conference was held at the Courant Institute on the macroscopic behavior of disordered media. We are grateful to the Exxon Research Corporation for the grant which made this conference possible.

The present volume contains most of the papers presented at this conference.

Research work on the macroscopic behavior of disordered media is very broad, ranging from applied engineering and experimental studies to more basic physics. It covers such diverse areas as numerical simulations, the numerical solution of phenomenological macroscopic equations, and mathematical investigations of the validity of macroscopic equations obtained from microscopic laws. In order to facilitate communication between the participants and to limit the length of the conference we decided to focus attention on the passage from microscopic to macroscopic laws and associated mathematical problems. We therefore excluded experimental and numerical studies. However, even within the areas selected, we left many problems such as localization, surface effects, fluid mixtures, etc., unrepresented.

During the course of the conference it became clear that there is a significant concentration of scientists in the New York metropolitan area actively engaged in research on macroscopic effects. Professor Morrel Cohen suggested that perhaps an informal conference among specialists take place annually (or more frequently) in order to exchange ideas and identify problems of theoretical and applied interest.

We expect to hold such an informal gathering at the Courant Institute in early summer 1982.

R. Burridge, S. Childress, G. Papanicolaou

ELECTROMAGNETIC PROPAGATION IN MIXED MEDIA

N. W. Ashcroft
Laboratory of Atomic and Solid State Physics
and
Materials Science Center
Cornell University, Clark Hall, Ithaca, NY 14853

I. Introduction

Colloids and other dilute aggregates of fine metal particles are frequently characterized by rather striking optical properties in the visible range of the spectrum. These and other interesting physical curiosities of such systems gave historical impetus to the study of an important form of condensed matter that, as we would now say, possesses microscopic order but it macroscopically disordered. By <u>fine</u> we imply a scale of length L that is small compared to the wavelength λ of the electromagnetic radiation probing the system. Actually in a mixed medium consisting of two constituents, each of which is characterized by a bulk (scalar) dielectric constant ϵ_i and magnetic permittivity μ_i, there are two intrinsic wave vectors $k_i = (\epsilon_i \mu_i)^{1/2} 2\pi/\lambda$, and five possible long wavelength conditions.[1] These are: $|k_1 L| \ll 1$, $|k_2 L| \ll 1$, $|kL| \ll 1$, $|k_1 b| \ll 1$, and $|kb| \ll 1$. Here k is the wave vector for the composite medium and is generally complex. The last two conditions are introduced to reflect the common experimental circumstance that in composite media physical clustering or other local ordering often takes place. The length scale characterizing such ordering is taken to be b. An important example is the periodic dielectric in which identical structures of constituent 2 (say) are embedded in crystalline fashion in a host of constituent 1. Here b will be taken as the magnitude of the largest primitive lattice vector. When the physical circumstances are such that all 5 conditions are met, the system is said to be in the quasistatic or infinite wavelength limit, and in this limit it can often be described by a simple average dielectric constant (or average permittivity). Otherwise the quantity of direct physical importance is the average product $(\epsilon\mu)_{av}$, as will be seen below.

II. Effective Medium Theories

The spatial arrangement of the constituents in the mixture will certainly be reflected in the manner in which boundary charges and local fields are established. In particular the connectedness of the composite has a great deal to say about the structure of the approximate theory used to describe the system. This can be demonstrated most easily for the two-component case by considering the mean field description of systems conforming to two physically important geometries. We shall first let the composite be divided into <u>cells</u>, each of which we imagine to be embedded in an effective medium of dielectric constant $\bar{\epsilon}$ (we take $\mu_1 = \mu_2 = 1$ for the present). Then in the quasi-static limit just described, there is an implicit averaging whose effect is to render the composite to be both translationally and rotationally invariant. Accordingly as a first approximation it is plausible to replace each irregularly shaped individual cell by spherical average cells. The question of interest is now this: how shall $\bar{\epsilon}$ of the effective medium be chosen to ensure that scattering of radiation from the average cell actually vanishes? The attributes of connectedness now enter in dictating the nature of those average cells. Consider first a medium in which particles of constituent 2 are always surrounded by constituent 1 and that the fractional occupancy of constituent 2 is η. In this example there is a single average cell. It is a coated sphere (cs) which, though properly to be regarded in its own right as a composite body (constituent 2 coated with constituent 1), can nevertheless be replaced in the long wavelength limit by a <u>uniform</u> sphere with dielectric constant

$$\bar{\epsilon}_{cs} = \epsilon_1 \left[\frac{2\epsilon_1 + \epsilon_2 + 2\eta(\epsilon_2 - \epsilon_1)}{2\epsilon_1 + \epsilon_2 - \eta(\epsilon_2 - \epsilon_1)} \right]$$

It follows that if the coated sphere is embedded in an effective medium with dielectric constant $\bar{\epsilon}$ chosen to be

$$\bar{\epsilon} = \epsilon_1 \left[\frac{2\epsilon_1 + \epsilon_2 + 2\eta(\epsilon_2 - \epsilon_1)}{2\epsilon_1 + \epsilon_2 - \eta(\epsilon_2 - \epsilon_1)} \right] \qquad (1)$$

then the no-scattering condition is trivially satisfied. This result can also be derived by standard Lorentz local field arguments.[2] Note that (1) can also be rewritten as

$$\bar{\epsilon} = \epsilon_1 \left[\frac{\epsilon_1 + (P(1-\eta) + \eta)(\epsilon_2 - \epsilon_1)}{\epsilon_1 + P(1-\eta)(\epsilon_2 - \epsilon_1)} \right] \qquad (2)$$

where P is the depolarizing factor which, for this spherical geometry, is 1/3. Equation (1) is the well known Maxwell-Garnett (MG) result[2,3] and is appropriate to systems conforming to the cermet geometry.

Apart from the basic topological statement requiring regions of 2 <u>always</u> to be surrounded by regions of 1, it contains no reference to microscopic features of the structure of the system. It is a particularly interesting result because irrespective of such details, and indeed, irrespective of connectedness, it can be shown[4] that (1) together with the complement formed by interchange of 1 and 2, form a pair of bounds on $\bar{\epsilon}$ when ϵ_1 and ϵ_2 are real. The generalization of such bounds to complex ϵ_i has been given by Milton[5] and Bergman.[6]

A quite different arrangement of the constituents, and one that is also physically realizable, leads to a very different choice of average cell. In the <u>aggregate</u> topology the components are arranged in such a way that in terms of connectedness each occurs on a completely equal footing, so that in contrast to the cermet topology neither can be considered host. Then with the same implicit averaging procedures in mind, there must be two types of average cell to be embedded in the effective medium. Each cell is again a sphere: we therefore have two spheres of each constituent and the no-scattering condition is very easily implemented by relating the fields inside and outside of them according to

$$E_{in}/E_{out} = 3/(2 + \epsilon_{in}/\epsilon_{out}).$$

A volume average then leads to the equation

$$\frac{3(1 - \eta)}{2 + \epsilon_1/\bar{\epsilon}} + \frac{3\eta}{2 + \epsilon_2/\bar{\epsilon}} = 1 \tag{3}$$

whose solution gives $\bar{\epsilon}$. This result is originally due to Bruggeman[7]: it is also widely known as the effective medium result, though both arguments just given are effective medium constructs. It will be noted that the symmetry present in the specification of the spatial arrangement of the aggregate topology is reflected in the symmetric occurrence of ϵ_1 and ϵ_2 in (3). Otherwise, however, there is no further reference to the microscopic arrangement of material in the system.

A real composite system must ultimately be described statistically through the specifications of correlation functions that specify the average microscopic arrangement of the constituents. There are correlations in such systems that go far beyond the gross connectedness properties just described. For example, in a dense composite we surely expect clustering effects in which regions of component 2 will be wholly surrounded by regions of component 1, and vice versa. It is possible to partially amalgamate (2) and (3) to reflect these correlations by appealing to an assumption of statistical independence in the occurrence of such regions. For a region in which 2 is enveloped by 1 we have

$$\bar{\epsilon}_1(\eta) = \epsilon_1 \left[\frac{2\epsilon_1 + \epsilon_2 + 2\eta(\epsilon_2 - \epsilon_1)}{2\epsilon_1 + \epsilon_2 - \eta(\epsilon_2 - \epsilon_1)} \right]$$

while for the opposite case

$$\bar{\epsilon}_2(\eta) = \epsilon_2 \left[\frac{2\epsilon_2 + \epsilon_1 + 2(1-\ )(\epsilon_1 - \epsilon_2)}{2\epsilon_2 + \epsilon_1 - \ (\epsilon_1 - \epsilon_2)} \right]$$

If the composite is regarded as an aggregate of such regions with relative occupancy preserved, then the average dielectric function is given by the solution to

$$\frac{3(1-\bar{\eta})}{2 + \bar{\epsilon}_1(\eta)/\epsilon} + \frac{3\bar{\eta}}{2 + \bar{\epsilon}_2(\eta)/\epsilon} = 1 \qquad (4)$$

The proper determination of $\bar{\epsilon}_1$ and $\bar{\epsilon}_2$ is dictated by microscopic considerations reflecting the structure of the system. An approach that takes a point of view similar to this has recently been given by Sheng.[8] The point to note is that the analytic structure of the expressions determining ϵ can be changed substantially by the further imposition of physical constraints. In mean field arguments such as those just given many of the details of the structure are obliterated by the averaging procedures. Such effects as are preserved are mainly the manifestations of dipolar fields. To incorporate structural multipole effects it is necessary to develop a theory that includes a more detailed treatment of correlations in such systems.

III. Multiple Scattering Methods

An example of a highly correlated mixed medium is a periodic array of identical inclusions (type 2, say) in a host of type 1. This crystalline arrangement falls within the definition of the cermet topology, as described above. If the inclusions have radius a, and a mean number density $\rho = N/V$, then Maxwell's equations lead to a wave equation for, say $\vec{H}(\vec{r})$, namely

$$\vec{\nabla} \times (\vec{\nabla} \times \vec{H}(\vec{r})) - (\omega^2/c^2)\epsilon(\vec{r})\mu(\vec{r})\vec{H}(\vec{r}) - \epsilon^{-1}(\vec{r})(\vec{\nabla}\epsilon(\vec{r})) \times (\vec{\nabla} \times \vec{H}(\vec{r})) = 0$$

(5)

where $\epsilon(\omega,\vec{r}) = \epsilon_1(\vec{r})$ for $\vec{r}$ in the host, and ϵ_2 otherwise (similarly for μ). There is a complementary equation for $\vec{E}$. Equation (5) can be regarded as a vector-field analog of the corresponding scalar field problem familiar from the context of one-electron band structure in crystalline systems. In the latter we are required to solve a Schrödinger equation in which the potential is periodic. The existence of a group of translations allows the solutions for the energy E to be indexed by a wave-vector $\vec{k}$, the Bloch wave-vector, which is restricted to a primitive cell of reciprocal space (the first Brillouin zone). The energy is then a function of the quasi continuous $\vec{k}$; the corresponding bands are bounded below and near their minima are quadratic in $\vec{k}$. Indeed for systems with sufficiently high symmetry, the energy for small k mimics that expected of a free particle but with an effective mass (rather than the free particle mass) which is the sole reminder of interactions that the electron has with the system. There is an immediate parallel with the periodic dielectric: again the translation symmetry allows us to index the solutions for $\vec{H}$ by a Bloch wave-vector. For a given frequency ω we shall seek solutions for wave-vectors of the form $\vec{k} = \vec{k}(\omega)$. In the long wavelength limit we shall then define an effective product $(\epsilon\mu)_{av}$ by the relation

$$(\epsilon\mu)_{av} = k^2(\omega)c^2/\omega^2 \qquad (k \to 0) \qquad (6)$$

as appropriate to plane waves.

The solution of equation (5) (or its complement for $\vec{E}$) is obtained from a variant[1] on the Green's function or Korringa-Kohn-Rostoker method.[9-11] A synopsis of this variant is now described: full details may be found in Ref. 1. The equation for $\vec{H}$ is rewritten as an integral equation

$$\vec{H}(\vec{r}') = -\int_S ds\{[\hat{n}\cdot\vec{\nabla}G_{\vec{k},\kappa}(\vec{r}',\vec{r})]\vec{H}(\vec{r}) - G_{\vec{k},\kappa}(\vec{r}',\vec{r})\hat{n}\cdot\vec{\nabla}\vec{H}(\vec{r})\} \qquad (7)$$

with $\kappa = (\epsilon\mu_1)^{1/2}\omega/c$, and the surface S is taken to be the primitive cell boundary and a surface infinitesimally outside of the spherical inclusion. Further, $\hat{n}$ is a normal that is outward and inward on cell and inclusion, respectively. The quantity G is defined by

$$G_{\vec{k},\kappa}(\vec{r}',\vec{r}) = \frac{1}{4\pi}\sum_{\vec{R}}\frac{\exp i\kappa|\vec{r}-\vec{r}'-\vec{R}|\exp i\vec{k}\cdot\vec{R}}{|\vec{r}'-\vec{r}-\vec{R}|} \qquad (8)$$

and is the Green's function (the centers of the spherical inclusions being located by the set of lattice vectors $\vec{R}$). Equation (7) is further reduced by performing an expansion of $\vec{H}(\vec{r})$ in electric and magnetic multipoles:

$$H(\vec{r}) = \vec{H}_E(\vec{r}) - \frac{ic}{\omega\mu} \vec{\nabla} \times \vec{E}_H(\vec{r}) \tag{9}$$

where

$$\vec{H}_E(\vec{r}) = \sum_{\ell m} A^E_{\ell m} f^E_\ell(r) \vec{L} Y_{\ell m}(\hat{r}) \tag{10}$$

with

$$\vec{L} = - i\vec{r} \times \vec{\nabla}$$

The terms in (10) are the electric multipoles. Equation (9) (and its complement) automatically satisfies the Maxwell divergence equations: the other two are satisfied by insisting that (10) also satisfies the wave equation. As noted in Ref. 1, the radial function $f^E_\ell(r)$ then satisfies a radial equation whose solutions for the spherical inclusion problem can be stated most succinctly in terms of a set of phase shifts. The details of the arrangement of these inclusions actually reside in the specification of the set of vectors $\{\vec{R}\}$, and hence in the Green's function G which has itself a standard expansion when $r' < r$. The expansion coefficients $A^E_{\ell m}$ can therefore be determined from the set of equations that emerge when the expanded quantities ($\vec{H}$ and G) are substituted into equation (7) and the result integrated over the surface of the spherical inclusion.

The solution of this secular problem leads via (6) in the long-wavelength limit to an effective product $(\epsilon\mu)_{av}$ which in general does not further separate. However, if the expansion for $\vec{H}$ (or $\vec{E}$) is truncated at $\ell = 1$ (dipoles) then for small ka it can be shown[1] that

$$(\epsilon\mu)_{av} = \epsilon_{MG}\mu_{MG} \tag{11}$$

i.e., <u>the product of Maxwell-Garnett results</u>[12] of the form (1). This verifies that the effective medium argument as applied to the cermet topology includes only dipolar effects: the result is obtained from a multiple scattering theory, and it is clear that structural effects will be included by retaining terms in (10) beyond $\ell = 1$. This is

done by using matrix folding techniques on the secular equation
determining the relation between k and ω. (The convergence of this
procedure is generally assured providing $Ka \ll 1$ and $k_2 a \ll 1$. That
the latter can be easily violated in metallic systems simply means
that eddy currents, i.e., magnetic multipoles, can certainly be
important.) The result is an expression for the average product that
is again separable (see equation (11)) in which the dielectric constant
ϵ has the form given in equation (2). In this case, however, the
depolarizing factor is no longer that appropriate to a sphere: it is
modified by the presence of structurally induced multipoles. The
latter now depend quite explicitly on the structure of the system and
can give rise to multipole resonances in physical observables whenever
the condition

$$\frac{\epsilon_2(\omega) - \epsilon_1}{(\ell + 1)\epsilon_1 + \ell\epsilon_2(\omega)} \approx 0 \qquad (\ell > 1)$$

is satisfied quasistatically. Physically these resonances relate to
the higher excited modes of the free dielectric inclusion.

Though derived from a theory devised to describe a periodic
array of dielectric inclusions, it is clear that (11) makes no reference to either the arrangement or the size of the constituents. It
therefore may be anticipated that a straightforward modification of
the theory can be constructed that will also hold true as the lowest
order solution for an aperiodic system with the cermet topology.
As shown in Ref. 1, this is in fact the case. The required modification is to take a macroscopic sample of aperiodic cermet, to periodically extend it, and to regard the resulting <u>grand crystal</u> as a
Bravais lattice with an exceedingly large basis. Though notationally
very cumbersome, the method just described is equally applicable to a
system with a basis: it resembles closely the associated band structure problem for electrons moving in a crystal described as a lattice
with a basis. Accordingly, in demanding a non-trivial solution for
the multipole expansion coefficients, a secular equation again results,
the principal difference from the previous case being that the matrix
elements are distinguished according to site. Again, truncation at
$\ell = 1$ gives the simplest results. The effective product $(\epsilon\mu)_{av}$ is
separable, and the average $\bar{\epsilon}$ has the Maxwell-Garnett form. Multipole
effects can again be included by matrix folding techniques and the
result is that provided that $Ka \ll 1$ and $k_2 a \ll 1$ there is a product
rule[12] and an expression for $\bar{\epsilon}$ similar to (2) in which the

depolarizing factor is modified by structural multipoles which now reveal the disorder in the system. Indeed, the determination of these corrections requires structural information normally provided by the pair and higher correlation functions for the inclusions. As shown in Ref. 1, the multiple scattering approach can be extended still further to include the fact that the sizes of particles in the system may also be dispersed. To handle this case it is only necessary to assume that the size of a given inclusion is uncorrelated with its spatial position, an assertion that is physically plausible. For this situation also, the results just described for the monodispersed case continue to hold with minor modifications (not surprisingly the form of the structural multipole modifications are somewhat more complex).

Thus, at least for mixed media distributed with the cermet topology, it is possible systematically to improve the mean field treatments of their electromagnetic properties. To do so, however, it is necessary to have access to structural information on the two-point, three-point, and higher correlation functions of the inclusions. In principle, it is possible to obtain such information on <u>pair</u> distributions from scattering experiments. However, for triplet functions and beyond it is necessary at present to resort to theoretical models. Ideally this structural information would also reveal the degree to which clustering and other local aberrations from average disorder are present. The occurrence of such regions leads, as noted earlier, to other length scales and hence to other conditions that must be satisfied in order to physically approach the long-wavelength and quasi-static limits.

Finally it should be noted that while the methods just described are suitable for the cermet topology they are clearly not applicable to the aggregate topology. In the electronic analog--the disordered alloy problem--the Bruggeman approach is related to the coherent potential approximation (the correspondence in the case of Maxwell-Garnett is to the average t-matrix approximation). In multiple scattering approaches to the alloy problem it is known[13] that some correlation effects can be included, albeit approximately, which suggests that provided there is adequate structural information, analogous procedures may be used here. For the periodic version of the aggregate topology it is certainly possible to use standard plane wave expansions in an attempt to solve (5). The intent is again to obtain the dispersion $\omega = \omega(\vec{k})$ and thence via (6) to an effective product $(\epsilon\mu)_{av}$. But on physical grounds and from experience with the electronic case, the convergence of such a procedure can only be expected to be satisfactory when the difference $\epsilon_2 - \epsilon_1$ is small compared with ϵ_1 and ϵ_2.

Acknowledgments

This work has been supported in part by the Solar Energy Research Institute (Grant #XH-9-8158-1) and in part by the National Science Foundation through the facilities of the Materials Science Center at Cornell University (Grant #DMR70-24008-A02). It is largely based on Ref. 1 toward which my co-authors Drs. Lamb and Wood made major contributions.

References and Footnotes

1. W. Lamb, D. M. Wood and N. W. Ashcroft, Phys. Rev. B<u>21</u>, 2248 (1980).

2. R. Landauer, Proceedings of the First Conference on the Electrical and Optical Properties of Inhomogeneous Media, Ohio State University, 1977, edited by J. C. Garland and D. B. Tanner (AIP, New York, 1978), No. 40, p. 2.

3. J. C. Maxwell-Garnett, Philos. Trans. R. Soc. London <u>203</u>, 385 (1904).

4. W. Brown, Jr., J. Chem. Phys. <u>23</u>, 1514 (1955); Z. Hashin and S. Shtrikman, J. Appl. Phys. <u>33</u>, 3125 (1962).

5. G. W. Milton, Appl. Phys. Letts. <u>37</u>, 300 (1980).

6. D. Bergman, Phys. Rev. Letts. <u>44</u>, 1285 (1980).

7. D.A.G. Bruggeman, Ann. Phys. (Leipzig) <u>24</u>, 636 (1935).

8. P. Sheng, Phys. Rev. Letts. <u>45</u>, 60 (1980).

9. J. Korringa, Physica (Utrecht) <u>13</u>, 392 (1947).

10. W. Kohn and N. Rostoker, Phys. Rev. <u>94</u>, 1111 (1954).

11. P. M. Morse, Proc. Nat. Acad. Sci. (USA) <u>42</u>, 276 (1956).

12. C. G. Granqvist, Z. Phys. B<u>30</u>, 29 (1978).

13. R. J. Elliot, J. A. Krumhansl and P. L. Leath, Rev. Mod. Phys. <u>46</u>, 465 (1974).

RESONANCES IN THE BULK PROPERTIES OF
COMPOSITE MEDIA - THEORY AND APPLICATIONS*

by

David J. Bergman
Department of Physics and Astronomy
Tel-Aviv University
69978 Tel-Aviv, Israel[†]

and

Department of Physics
The Ohio State University
Columbus, Ohio 43210

Contents
I. Introduction
II. Electrostatic Problems - General Theory
III. Rigorous Bounds for Real and Complex ϵ_e in a Two-Component Composite
IV. Calculation of the Electrostatic Resonances of a Two-Component Composite
V. Electrostatic Problems in Multicomponent Composites
VI. Extension of the Approach to Other than Electrostatic Problems
VII. Conductivity or Connectivity Thresholds
VIII. Acknowledgements and Bibliography

I. INTRODUCTION

A composite material is an inhomogeneous medium where the scale of the inhomogeneities is macroscopic. There are small regions in which the properties are uniform and equal to the bulk properties of one of the components. On larger scales, the composite can usually be described by a set of effective bulk properties, as though it too were uniform. A central problem in the theory of composites is the calculation of such bulk properties. The main difficulty arises when the bulk property depends on the detailed microgeometry of the composite. For a complete solution, one needs to calculate the physical fields inside the composite, including all their distortions caused by the inhomogeneous structure. Even when the microgeometry is known precisely, this is a very difficult task that has usually been approached by attempting to solve numerically the partial differential equations for the local fields, including continuity conditions at the interfaces between the different pure components.

We note here that even when the composite is made of separate, simply shaped inclusions (e.g., spheres) of one pure material embedded in another pure material, a

---

*Research supported in part by the United States - Israel Binational Science Foundation under Grant No. 2006/79.

[†]Permanent address

calculation along these lines does not simplify. That is because, unless the density of inclusions is very low, or they differ very little from the host material in their material property, the distortive effects of different inclusions are not additive. Consequently the solution of the problem for an isolated spherical inclusion does not help us, and we must consider all of them simultaneously. This is similar to the case of a multiple scattering problem, where the field impinging on any scatterer includes the scattered field from all the other scatterers.

An entirely different approach to such problems, which we will advocate here, is to focus attention on the resonance states of the problem. These resonances are eigenfunction solutions of the same equation that determines the local field in the composite in the presence of an external field, but satisfy different boundary conditions on the exterior surface of the system (i.e., no external field is present). It is a very natural idea to try to expand the local field of the real problem (i.e., the one where there is an external field present) in these eigenfunctions. In this way, a very convenient representation is obtained both for the field and for other interesting quantities, in particular for the effective bulk material properties of the composite.

Some advantages of this type of representation are: a) The resonances depend on the microgeometry, but they are independent of the specific material properties of the inclusions, and in some cases (e.g., the dielectric constant) they are also independent of the host properties. Once they are known, they provide us with a solution for any type of inclusion that has the same microgeometry. b) When the microgeometry is only partially known, the positions and strengths of the resonances can be treated as free parameters that are varied in order to find rigorous upper and lower bounds on the effective bulk properties. This turns out to be a very efficient procedure for obtaining such bounds. c) When the microgeometry is precisely known, the resonances can often be calculated in stages. If the inclusions can be described as made up of simply shaped grains (e.g., spheres that do or do not overlap), then we can first find the resonances of each grain by treating it as an isolated inclusion. The totality of these eigenstates from all the grains are then used as a basis in the function space on which to set up the problem for the entire composite. The resonances of the entire composite are then the eigenvectors of an infinite matrix whose diagonal elements are the individual grain eigenvalues, while the off-diagonal elements are overlap integrals between eigenstates of two different grains. This means that instead of having to solve a multiple scattering problem for the actual local field, we only have to solve a problem of two-body interactions to find the resonances. This approach enables us to take the utmost advantage of any symmetries of the individual grain (e.g., if the grains are spherical the isolated grain eigenfunctions and even the two-body overlap integrals are very easy to obtain) even when the total symmetry of the composite is very low. If the composite has a periodic structure, a Bloch theorem can be shown to hold for the eigenfunctions and a further simplification results. d) Since the resonances are the singular points of the effective bulk property as a function of the material properties of the pure components, they completely determine the analytical properties of that function. By considering the resonance spectrum, it is possible to make some rather general statements about particular singularities of that function. For example, by considering a system or an ensemble of systems that has a conductivity threshold at some value of a continuously varying parameter, we can predict singular behavior at that point for a variety of other physical properties of the system.

To date, this approach has been applied to electrostatic problems (Bergman, 1978 a,b, 1979 a,b), to finite frequency, quasi-electrostatic problems (Bergman, 1980, 1981, a,b), to random-resistor-networks (Bergman and Kantor, 1981), to elastostatic problems (Kantor and Bergman, 1981 a,c), and to the scattering of electromagnetic waves (Bergman and Stroud, 1980). An application to quantum mechanical scattering and to other classical problems in composite systems are under consideration. Since the method was first proposed in the context of electrostatic problems, and since that is the field where the most applications have been made, we will mostly discuss those problems here, pointing out where differences arise in the discussion of other types of problems.

II. ELECTROSTATIC PROBLEMS - GENERAL THEORY

The basic properties that are crucial to all later developments are: a) For an $(m+1)$-component composite, with pure component dielectric constants $\epsilon_1 \ldots \epsilon_{m+1}$, the effective dielectric constant $\epsilon_e(\epsilon_1 \ldots \epsilon_{m+1})$ is an analytic function in most of the complex space of $\epsilon_1 \ldots \epsilon_{m+1}$. b) The function $\epsilon_e(\epsilon_1 \ldots \epsilon_{m+1})$ is homogeneous of order one, and consequently we can eliminate its dependence on any single component (called the "host material"), e.g., ϵ_{m+1}.

These properties can be demonstrated in different ways. We will choose a method that is useful also for further practical developments. The problem of calculating the electrostatic (e.s.) field ϕ in a composite that lies between the plates of a large parallel plate condenser is set up as an integral equation (compare with Bergman, 1979 a,b)

$$\phi = z + \sum_{i=1}^{m} u_i \Gamma_i \phi$$

$$\Gamma_i \phi \equiv \int dV' \, \theta_i(\vec{r}') \, \frac{\partial G(\vec{r},\vec{r}')}{\partial \vec{r}'} \cdot \frac{\partial \phi(\vec{r}')}{\partial \vec{r}'}$$

$$\theta_i(\vec{r}) = \begin{cases} 1 & \text{for } \vec{r} \text{ inside } \epsilon_i \text{ material} \\ 0 & \text{otherwise} \end{cases}$$

$$u_i \equiv \frac{1}{s(i)} \equiv 1 - \frac{\epsilon_i}{\epsilon_{m+1}} \quad . \tag{II.1}$$

Here $G(\vec{r},\vec{r}')$ is Green's function for the parallel plate condenser

$$\nabla^2 G(\vec{r},\vec{r}') = -\delta^3(\vec{r}-\vec{r}')$$

$$G = 0 \text{ for } \vec{r} \text{ on the plates}$$

$$\frac{\partial G}{\partial n} = 0 \text{ for } \vec{r} \text{ on the walls,} \tag{II.2}$$

and for an infinitely large condenser it is equal to the Coulomb potential of a point charge

$$G(\vec{r},\vec{r}') = \frac{1}{4\pi|\vec{r}-\vec{r}'|} . \tag{II.3}$$

The inhomogeneous term z in (II.1) takes care of the boundary condition on ϕ that arises from assuming that the condenser plates are perpendicular to the z-axis, and that a potential difference is applied that is equal (numerically) to the distance between them. (This means that if the condenser were filled with a homogeneous dielectric, the electric field would be 1.) The linear integral operator Γ_i, which uses values of ϕ only inside ε_i material, is hermitian under the following definition of a scalar product

$$(\psi,\phi)_i \equiv \int dV\, \theta_i\, \vec{\nabla}\psi^* \cdot \vec{\nabla}\phi . \tag{II.4}$$

Restricting ourselves, for the time being, to a two-component composite, it is easy to convince ourselves that the eigenvalues s_n of $\Gamma(\equiv\Gamma_1)$ must all lie on the real axis between 0 and 1. Any other (real) value for an eigenvalue s_n would mean that there exists no solution for the electrostatic potential in a charged capacitor filled with a composite made of two materials with certain real, positive values of $\varepsilon_1, \varepsilon_2$ - clearly an impossibility.

The (differentiable) functions in the ε_1-volume form a Hilbert space under the scalar product of (II.4), and we can use the eigenfunctions of Γ as a basis set, but only inside the ε_1-volume. Using the usual expansion for the unity-operator, and introducing a convenient bra-ket notation, we get the following equation for the expansion coefficients of the potential function ϕ

$$\langle\phi_n|\phi\rangle = \langle\phi_n|z\rangle + \sum_m \langle\phi_n|u\Gamma|\phi_m\rangle\langle\phi_m|\phi\rangle . \tag{II.5}$$

From this we get

$$\langle\phi_n|\phi\rangle = \frac{\langle\phi_n|z\rangle}{1 - u s_n} \tag{II.6}$$

and hence, inside the ε_1-volume, the following expansion for ϕ

$$|\phi\rangle = \sum_n |\phi_n\rangle \frac{\langle\phi_n|z\rangle}{1 - u s_n} . \tag{II.7}$$

In order to get an expansion for ϕ that is valid also outside the ε_1-volume, we substitute (II.7) on the right hand side of the integral equation for ϕ, thus getting

$$|\phi\rangle = |z\rangle + u\,\Gamma|\phi\rangle = |z\rangle + \sum_n |\phi_n\rangle \frac{s_n \langle \phi_n|z\rangle}{s - s_n} \quad , \tag{II.8}$$

where we have introduced the new variable

$$s \equiv \frac{1}{u} = \frac{\epsilon_2}{\epsilon_2 - \epsilon_1} \quad . \tag{II.9}$$

The bulk effective dielectric constant of the composite ϵ_e can be defined by requiring that if the capacitor were filled by a fictitious homogeneous dielectric ϵ_e, the (uniform) displacement field would be equal to the z-component of the volume averaged displacement field in the inhomogeneous sample

$$\epsilon_e \equiv \frac{1}{V} \int \epsilon(\vec{r})\, \frac{\partial \phi}{\partial z}\, dV \quad . \tag{II.10}$$

Alternatively, one could require the electrostatic energy stored in the condenser to be reproduced correctly - the two definitions can easily be shown to be equivalent. Noting that we can represent the local dielectric constant $\epsilon(\vec{r})$ with the help of the θ-functions as follows

$$\epsilon(\vec{r}) = \epsilon_2(1 - u\,\theta_1(\vec{r})) \quad , \tag{II.11}$$

we can transform (II.10) into

$$F(s) \equiv 1 - \frac{\epsilon_e}{\epsilon_2} = \frac{u}{V} \int dV\, \theta_1 \frac{\partial \phi}{\partial z} = \frac{u}{V}\langle z|\phi\rangle = \sum_n \frac{F_n}{s - s_n} \tag{II.12a}$$

$$F_n \equiv \frac{1}{V}|\langle z|\phi_n\rangle|^2 \quad , \tag{II.12b}$$

where we have now introduced and defined one of the "characteristic geometric functions" of the composite, $F(s)$. This is the basic representation for ϵ_e of a two-component composite in terms of its resonances.

As we noted previously, the poles s_n (i.e., eigenvalues of Γ) are all between 0 and 1. Due to completeness of the set ϕ_n, the weights F_n obey the following sum rule

$$\sum_n F_n = \frac{1}{V} \sum_n \langle z|\phi_n\rangle\langle\phi_n|z\rangle = \frac{1}{V}\langle z|z\rangle = p_1 \quad , \tag{II.13}$$

where p_1 is the volume fraction of ϵ_1 material. Since the weights are all positive, it obviously follows that they must all lie between 0 and p_1. It can be shown that, for a composite with isotropic or cubic point symmetry, another sum rule holds, namely,

$$\sum_n F_n s_n = \frac{1}{3} p_1 p_2 \tag{II.14}$$

where $p_2 = 1 - p_1$ is the volume fraction of ϵ_2 (see Bergman, 1978 a,b). The higher moments of the pole spectrum

$$\sum_n F_n s_n^r \quad , \quad r \geq 2 \tag{II.15}$$

depend on $(r+1)$-order correlation functions of the form

$$\theta_1(\vec{r}_1)\ \theta_1(\vec{r}_2)\ldots\theta(\vec{r}_{r+1}) \ . \tag{II.16}$$

A knowledge of these moments can be used to generate an expansion for $F(s)$ in powers of $u = 1/s$:

$$F(s) = \sum_{r=0}^{\infty} \frac{\sum_n F_n s_n^r}{s^{r+1}}$$

$$= p_1 u + \frac{1}{3} p_1 p_2 u^2 + \sum_n F_n s_n^2 \cdot u^3 + \ldots \ . \tag{II.17}$$

A further property of the resonance expansion (II.12) that is worth emphasizing is that the poles s_n and the weights F_n are entirely determined by the microgeometry of the composite, and are totally independent of the physical properties of the pure components, i.e., ϵ_1 and ϵ_2. The values of the two dielectric constants enter only through the variable s. Consequently, a knowledge of the parameters s_n, F_n enables us to calculate ϵ_e for any two-component composite with the same microgeometry, and even other bulk properties, such as conductivity, magnetic permeability, diffusivity - since all these physical processes are described by the same mathematical structure as the electrostatic potential field in a dielectric.

Before concluding this discussion of the two-component composite system, I would like to point out that there have been published some incorrect statements about the distribution of singularities of the function $F(s)$ (see, e.g., Bergman 1978 b, p. 384). A careful analysis reveals that the qualitative features of this distribution depend on the singularities in the shape of the ϵ_1, ϵ_2 interface. In particular, for a finite system with a smooth interface (i.e., no corners and no isolated points or lines of contact), the spectrum of s_n is a discrete spectrum, with a single accumulation point at $s = \frac{1}{2}$ (Bergman 1981 c).

In the case of a multicomponent composite, (i.e., more than two pure components,) we can still discuss $\epsilon_e(\epsilon_1\ldots\epsilon_{m+1})$ in terms of a characteristic geometric function, defined by

$$F(s^{(1)}\ldots s^{(m)}) \equiv 1 - \frac{\epsilon_e}{\epsilon_{m+1}} \ , \quad m > 1 \ . \tag{II.18}$$

The analytical properties of F have not been worked out completely for this case. If the interfaces are all smooth and the system is finite, the singularities of F hopefully lie upon a set of $(m-1)$-dimensional hypersurfaces in the complex space of $s^{(1)}\ldots s^{(m)}$. It can be shown that, at any real singular point, at least one of the $s^{(i)}$ must always lie between 0 and 1 (see Bergman, 1978 b). In practice, the multicomponent system has always been dealt with, until now, by choosing an appropriate trajectory in the complex space of $s^{(1)}\ldots s^{(m)}$. E.g., in attempting to derive rigorous bounds for multicomponent systems, the useful approach to take is to find a trajectory

$s^{(i)}(s)$ that passes through all the important points in that space, such that $F(s^{(i)}(s))$ has a representation similar to (II.12a), together with appropriate sum rules (see Bergman 1978 b).

III. RIGOROUS BOUNDS FOR REAL AND COMPLEX ϵ_e IN A TWO-COMPONENT COMPOSITE

In principle, for a two-component composite with real values of ϵ_1, ϵ_2 one should start from (II.12a), treat the F_n and s_n as free parameters, and attempt to maximize or minimize $F(s)$ subject to various constraints - equalities such as the sum rules (II.13),(II.14), and inequalities such as

$$0 \leq s_n < 1$$

$$0 < F_n \leq p_1 . \tag{III.1}$$

In practice, it is often better to define new functions, with properties similar to those of $F(s)$, that incorporate some of these constraints (i.e., the sum rules) automatically.

The first bound is obtained by noting that, when both ϵ_1 and ϵ_2 are positive (i.e., true physical dielectrics) then s is real and lies outside the segment $(0,1)$. In that case we can write

$$F(s) = \sum_n \frac{F_n}{s - s_n} > \frac{\sum F_n}{s} = \frac{p_1}{s} , \quad \text{for } s \notin (0,1). \tag{III.2}$$

Translated into $\epsilon_1, \epsilon_2, \epsilon_e$ language this inequality becomes

$$\epsilon_e < p_1 \epsilon_1 + p_2 \epsilon_2 , \tag{III.3}$$

i.e., a well known inequality for composites (Wiener, 1912).

A similarly well known inequality involving reciprocal ϵ's, namely (Wiener, 1912),

$$\frac{1}{\epsilon_e} < \frac{p_1}{\epsilon_1} + \frac{p_2}{\epsilon_2} \tag{III.4}$$

can be obtained by considering the function

$$H(t) \equiv 1 - \frac{\epsilon_2}{\epsilon_e} = \frac{F(s)}{F(s) - 1}$$

$$t \equiv \frac{\epsilon_1}{\epsilon_1 - \epsilon_2} = 1 - s . \tag{III.5}$$

Starting from the properties of $F(s)$, it can be shown that $H(t)$ has very similar properties, namely (see Bergman 1981 a,b),

$$H(t) = \sum_n \frac{H_n}{t - t_n}$$

$$\sum H_n = p_1 \tag{III.6a}$$

$$\sum_n H_n t_n = \frac{2}{3} p_1 p_2$$

$$0 \le t_n < 1$$

$$0 < H_n \le p_1 \tag{III.6b}$$

(the only difference is that in the second sum rule, the factor 1/3 has been replaced by 2/3). Obviously, H(t) contains the same information as F(s), and therefore, it too is a "characteristic geometric function" of the composite. Repeating the procedure of (III.2) for H(t) we obtain

$$H(t) > \frac{p_1}{t} , \quad \text{for } t \notin (0,1) , \tag{III.7}$$

and when this is translated into $\epsilon_1, \epsilon_2, \epsilon_e$ language, the result is (III.4).

A better pair of bounds for real ϵ_e can be obtained if we know not only the volume fractions p_1, p_2, but also that the system is isotropic or that it is cubic. To achieve this, we must include the sum rule (II.14) in our considerations. This is most easily done by defining a new, auxiliary geometric function,

$$C(s) \equiv \frac{1}{p_1} - \frac{1}{s\,F(s)} . \tag{III.8}$$

It is a straightforward matter to show that C(s) has a structure similar to F(s) (see Bergman 1981 b), namely

$$C(s) = \sum_n \frac{C_n}{s - s_n'}$$

$$0 \le s_n' < 1$$

$$0 < C_n \le \frac{1}{3} \frac{p_2}{p_1} . \tag{III.9}$$

Furthermore, the zero moment sum rule for F(s), (II.13), has gotten translated into the requirement that $C(s) \to 0$ as $s \to \infty$, while the first moment sum rule for F(s) (II.14), has gotten translated into a zero moment sum rule for C(s)

$$\sum_n C_n = \frac{1}{3} \frac{p_2}{p_1} . \tag{III.10}$$

If we now apply the procedure of (III.2) to C(s), we get

$$C(s) > \frac{p_2}{3 p_1 s} . \tag{III.11}$$

Translated back to F(s) and ϵ_e, this becomes

$$F(s) \gtrless \frac{p_1}{s - \frac{1}{3} p_2} \quad \text{for } s \gtrless \frac{1}{0} \tag{III.12a}$$

$$\epsilon_e \gtrless \epsilon_2 + \frac{p_1}{\frac{1}{\epsilon_1-\epsilon_2}+\frac{p_2}{3\epsilon_2}} \quad \text{for} \quad \epsilon_1 \lessgtr \epsilon_2 \quad . \tag{III.12b}$$

The last form is nothing but the well known pair of bounds discovered by Hashin and Shtrikman (1962).

In order to improve the bounds still further, more information is necessary. Such information could come in the form of higher moments of the pole spectrum s_n. Those are in general very difficult to obtain for systems with arbitrary microgeometries. It is however possible to obtain them for some simple, exactly known geometries, e.g., cubic arrays of spherical inclusions. This approach has been used by McPhedran and Milton (1981) to obtain a series of successively narrower pairs of bounds on ϵ_e for such systems as an alternative to the direct calculation of ϵ_e, such as we will describe in Sec. IV below.

A different way to include additional information in our considerations arises when we know the value of F at some other value of its argument - say s^+. This can happen if someone has measured a different physical property in the same composite, or in a composite with the same microgeometry. It can even happen if someone has measured the same property, e.g., ϵ_e, but at a different temperature or frequency, where ϵ_1, ϵ_2 and hence s have different values (see Prager 1969).

A knowledge of $F(s^+)$ leads to a further constraint on the variation of s_n, F_n of the form

$$F(s^+) = \sum_n \frac{F_n}{s^+ - s_n} \quad . \tag{III.13}$$

The best way to treat this case is again to define a new auxiliary geometric function

$$B(s) \equiv \frac{3}{p_1 p_2} - \frac{1}{s^2 F(s) - s p_1} \quad , \tag{III.14}$$

which can be shown to have a form similar to that of F(s), namely (see Bergman 1981 b)

$$B(s) = \sum_n \frac{B_n}{s - s_n''}$$

$$s_o'' = 0, \quad 0 < s_n'' < 1 \quad \text{for} \quad n \neq o$$

$$0 < B_n \quad . \tag{III.15}$$

In this case, both the zero moment and the first moment sum rules of F(s) are automatically incorporated in the definition of B(s), and we only have to worry about the new constraint (III.13). A straightforward calculation leads to

$$\frac{s B(s) - s^+ B(s^+)}{s - s^+} = - \sum_{n \neq o} \frac{B_n s_n}{(s - s_n)(s^+ - s_n)} \leq 0 \quad , \tag{III.16a}$$

$$\text{for } s, s^+ \text{ on } \begin{Bmatrix} \text{the same side} \\ \text{opposite sides} \end{Bmatrix} \text{ of } (0,1) \ . \tag{III.16b}$$

This gives us a bound for $B(s)$, and hence for $F(s)$ and ϵ_e.

To get the other bound, we start from another characteristic geometric function

$$E(s) \equiv 1 - \frac{\epsilon_1}{\epsilon_e} = \frac{1 - s\, F(s)}{s(1 - F(s))} \ , \tag{III.17}$$

which is similar to $H(t)$ but with the roles of ϵ_1, ϵ_2 reversed, and from it define an auxiliary geometric function similar to $B(s)$

$$B_E(s) = \frac{3}{2 p_1 p_2} - \frac{1}{s^2 E(s) - s\, p_2} \ . \tag{III.18}$$

This function can again be shown to have properties similar to $B(s)$. Applying to it the procedure of (III.16) we then obtain the other bound for $F(s)$ and ϵ_e.

Translated into $\epsilon_1, \epsilon_2, \epsilon_e$ language, the two bounds can be found by solving the following two equations for ϵ_e

$$\frac{\epsilon_2^+ - \epsilon_1^+}{\epsilon_e^+ - \langle \epsilon^+ \rangle} - \frac{\epsilon_2 - \epsilon_1}{\epsilon_e - \langle \epsilon \rangle} = \frac{3}{p_1 p_2} \frac{\epsilon_2^+ \epsilon_1 - \epsilon_1^+ \epsilon_2}{(\epsilon_2^+ - \epsilon_1^+)(\epsilon_2 - \epsilon_1)}$$

$$\frac{\tilde{\epsilon}_2^+ - \tilde{\epsilon}_1^+}{\tilde{\epsilon}_e^+ - \langle \tilde{\epsilon}^+ \rangle} - \frac{\tilde{\epsilon}_2 - \tilde{\epsilon}_1}{\tilde{\epsilon}_e - \langle \tilde{\epsilon} \rangle} = \frac{3}{2 p_1 p_2} \frac{\tilde{\epsilon}_2^+ \tilde{\epsilon}_1 - \tilde{\epsilon}_1^+ \tilde{\epsilon}_2}{(\tilde{\epsilon}_2^+ - \tilde{\epsilon}_1^+)(\tilde{\epsilon}_2 - \tilde{\epsilon}_1)} \ ,$$

$$\tilde{\epsilon} \equiv 1/\epsilon, \quad \langle \epsilon \rangle \equiv p_1 \epsilon_1 + p_2 \epsilon_2 \ . \tag{III.19a}$$

The first equation yields the upper (lower) bound and the second equation yields the lower (upper) bound on ϵ_e when the sign of

$$\frac{\epsilon_2^+ \epsilon_1 - \epsilon_1^+ \epsilon_2}{\epsilon_1 - \epsilon_2} \tag{III.19b}$$

is positive (negative). Obviously, we could go on and include more pieces of information of this type, and attempt to get even better bounds.

A further piece of information that has not been included until now, is an inequality that must hold for $\epsilon_e(\epsilon_1, \epsilon_2)$ and $\epsilon_e(\epsilon_2, \epsilon_1)$. The latter quantity represents a composite with the same microgeometry, but with the two components interchanged. For an isotropic or cubic composite, it can be shown that the following inequality must hold (see Schulgasser 1976, Bergman 1981 b)

$$\epsilon_e(\epsilon_1, \epsilon_2)\, \epsilon_e(\epsilon_2, \epsilon_1) \geq \epsilon_1 \epsilon_2, \quad \text{for } \epsilon_1, \epsilon_2 > 0 \ . \tag{III.20}$$

In order to include this too in our considerations, we define a new auxiliary function

$$A(s) \equiv \frac{s - \frac{1}{3} p_1}{p_1} - \frac{1}{F(s)} \quad . \tag{III.21}$$

This can again be shown to have the usual type of structure

$$A(s) = \sum \frac{A_n}{s - s_n'''}$$

$$0 < s_n''' < 1$$

$$0 < A_n \quad , \tag{III.22}$$

and the two sum rules on $F(s)$ are both automatically incorporated. Note that the poles of $A(s)$ are the same as the poles of $C(s)$ (i.e., zeros of $F(s)$), except for $s = 0$ which may be a pole of C but may not be a pole of A.

In terms of $A(s)$, (III.20) becomes

$$A(s) + A(1 - s) \leq \frac{1}{3} \frac{p_2}{p_1} \quad , \quad \text{for} \quad s \notin (0,1) \quad . \tag{III.23}$$

Further inequalities that $A(s)$ must satisfy for particular values of s are

$$A(0) + \frac{1}{3} \frac{p_2}{p_1} = -\frac{1}{F(0)} \geq 0 \tag{III.24}$$

$$A(1) - \frac{2}{3} \frac{p_2}{p_1} = 1 - \frac{1}{F(1)} \leq 0 \quad . \tag{III.25}$$

We note that (III.25) is also a consequence of (III.24) and of (III.23) taken at $s = 0$ or at $s = 1$.

Because (III.23) must hold for any s, it is not immediately clear how many independent constraints it is in fact equivalent to. For the time being we therefore restrict ourselves to considering only (III.23) at $s = 0$ (or at $s = 1$, which leads to the same inequality), and (III.24), thus

$$- A(0) \leq \frac{1}{3} \frac{p_2}{p_1} \tag{III.26}$$

$$A(0) + A(1) \leq \frac{1}{3} \frac{p_2}{p_1} \quad . \tag{III.27}$$

In addition to these inequalities, we must include the information about $A(s^+)$. In terms of the parameters A_n, s_n''' these three constraints become

$$\sum_n \frac{A_n}{s_n'''} \leq \frac{1}{3} \frac{p_2}{p_1} \tag{III.28}$$

$$\sum_n \frac{A_n (2 s_n - 1)}{s_n (1 - s_n)} \leq \frac{1}{3} \frac{p_2}{p_1} \tag{III.29}$$

$$\sum_n \frac{A_n}{s^+ - s_n} = A(s^+) \quad . \tag{III.30}$$

Note that these constraints, as well as the quantity we would like to maximize or minimize, $A(s)$, are all linear in the weights A_n. To these constraints we should also add the usual requirement that all $A_n \geq 0$. Each of these linear constraints defines a hyperplane in the space of the A_n.

According to linear programming theory (see, e.g., Dantzig 1963), an extremum point of $A(s)$ with respect to the A_n (the s_n''' are kept fixed for the time being) can only occur at a vertex of these hyperplanes. What this means is that most of the hyperplanes defining that vertex must be the trivial coordinate hyperplanes $A_n = 0$. Some of these trivial hyperplanes (at least one, at most three) can be replaced by the non-trivial hyperplanes of (III.28)-(III.30). Therefore we conclude that the bounding expression for $A(s)$ consists of at least one and at most three poles.

We now introduce another characteristic geometric function

$$G(t) \equiv 1 - \frac{\epsilon_e}{\epsilon_1} = \frac{1 - s\, F(s)}{1 - s}$$

$$t \equiv (1 - \frac{\epsilon_2}{\epsilon_1})^{-1} = 1 - s \quad , \tag{III.31}$$

which obviously is very similar to $F(s)$ except that the roles of ϵ_1 and ϵ_2 have been interchanged. By analogy with (III.21), we can define the auxiliary function

$$A_G(t) \equiv \frac{t - \frac{1}{3} P_1}{P_2} - \frac{1}{G(t)} \quad , \tag{III.32}$$

and show that the following relation holds between $A(s)$ and $A_G(t)$

$$\left(\frac{P_2}{P_1}(s - \frac{1}{3}) - A(s)\right)\left(\frac{P_1}{P_2}(t - \frac{1}{3}) - A_G(t)\right) = ts \quad . \tag{III.33}$$

We now consider the case where the bounding expression for $A(s)$ has three poles, and both (III.24) and (III.25) (as well as (III.27)) are satisfied as equalities. It is easy to see from (III.33) that in this case, $A_G(t)$ will have only two poles, and that it satisfies the analogues of (III.24) and (III.25) as strict inequalities

$$- A_G(0) < \frac{1}{3} \frac{P_1}{P_2} \tag{III.34}$$

$$A_G(1) < \frac{2}{3} \frac{P_1}{P_2} \quad . \tag{III.35}$$

From (III.33) we can also derive the following equation for $A(s) + A(t)$

$$\frac{1}{3}\frac{P_2}{P_1} - \left(A(s) + A(t)\right) = ts\left[\frac{1}{3}\frac{P_1}{P_2} - \left(A_G(s) + A_G(t)\right)\right]\left[\frac{P_1}{P_2}(t - \frac{1}{3}) - A_G(t)\right]^{-1}\left[\frac{P_1}{P_2}(s - \frac{1}{3}) - A_G(s)\right]^{-1} . \tag{III.36}$$

From this is follows that the analogue of (III.27) is satisfied as an equality

$$A_G(0) + A_G(1) = \frac{1}{3}\frac{p_1}{p_2} \ . \tag{III.37}$$

Thus, since $A_G(t)$ is as good a tool as $A(s)$ for constructing bounds on $F(s)$ or ε_e, we can always restrict ourselves to at most a two-pole bound for $A(s)$ or $A_G(t)$.

Returning now to consider $A(s)$, we assume a bounding expression that has two poles

$$\frac{A_0}{s-s_0} + \frac{A_1}{s-s_1} \ . \tag{III.38}$$

Solving (III.30) to get A_0 and substituting in (III.28) and (III.29), these two inequalities become, respectively,

$$A_1 \frac{s_0-s_1}{s^+-s_1}\frac{s^+}{s_0 s_1} \le \frac{1}{3}\frac{p_2}{p_1} - \frac{s^+-s_0}{s_0} A(s^+) \tag{III.39}$$

$$A_1 \frac{s_0-s_1}{s^+-s_1}\frac{s^+}{s_0 s_1} \ge \frac{1}{3}\frac{p_2}{p_1}\left(\frac{1-s^+}{(1-s_0)(1-s_1)}\frac{s_0 s_1}{s^+}-1\right)^{-1} - A(s^+)(s^+-s_0)\frac{s^+}{s_0 s_1} \ . \tag{III.40}$$

Using the same substitution in the bounding expression leads to the following form for the bound

$$\frac{s^+-s_0}{s-s_0} A(s^+) + A_1 \frac{(s_0-s_1)(s-s^+)}{(s-s_0)(s-s_1)(s^+-s_1)} \ . \tag{III.41}$$

This will be an upper (lower) bound if we maximize (minimize) it by varying A_1, s_0, s_1 subject to all the (inequality) constraints. At first, we will keep s_0, s_1 fixed and only allow A_1 to vary. In this process, one of the inequalities (III.39), (III.40) becomes an equality first, depending on the values of s, s^+, and on whether an upper or a lower bound is sought.

E.g., if $s > 1$, $s^+ < 0$, and we are looking for a lower bound, then we must increase the l.h.s. of (III.39) and (III.40) by changing A_1. Therefore (III.39) becomes an equality while (III.40) remains a strict inequality. Using (III.39) as an equality to solve for A_1 and substituting the result in (III.40) and (III.41) we obtain the following forms for the remaining inequality and bounding expression, respectively

$$A(s^+)\frac{(s^+-s_0)(s^+-s_1)}{s_0 s_1} + \frac{1}{3}\frac{p_2}{p_1}\frac{\dfrac{s^+}{1-s^+}\dfrac{(1-s_0)(1-s_1)}{s_0 s_1}-1}{\dfrac{s^+}{1-s^+}\dfrac{(1-s_0)(1-s_1)}{s_0 s_1}-1} \ge 0 \tag{III.42}$$

$$\frac{(s^+-s_0)(s^+-s_1)}{(s-s_0)(s-s_1)}\frac{s}{s^+}A(s^+) - \frac{1}{3}\frac{p_2}{p_1}\frac{s_0 s_1}{s^+}\frac{s-s^+}{(s-s_0)(s-s_1)} \ . \tag{III.43}$$

We must now still let s_0, s_1 vary to as to minimize (III.43). This can be done by decreasing either s_0 or s_1. Consequently, we can decrease s_0 and s_1 until (III.42) becomes an equality. The values of s_0, s_1 that minimize (III.43) will therefore satisfy (III.42) as an equality. Therefore, if the lower bound for $A(s)$ were a two-pole expression, then it would have to satisfy both (III.39) and (III.40) as equalities. Consequently, when translated into $A_G(t)$ language, the bounding function would become a one-pole expression $A_{Gh}(t) = A_0/(t - t_0)$ that satisfies only the analogue of (III.40) as an equality.

We now examine the possibility of a two-pole upper bound, still assuming that $s > 1, s^+ < 0$. We must now decrease the ℓ.h.s. of (III.39) and (III.40) as much as possible by changing A_1. Therefore (III.40) becomes an equality while (III.39) remains a strict inequality. Using (III.40) as an equality to determine A_1 and substituting the result in (III.39) and (III.41), we again obtain (III.42) for the remaining inequality, while the bounding expression becomes

$$\frac{(s^+ - s_0)(s^+ - s_1)}{(s - s_0)(s - s_1)} A(s^+) + \frac{1}{3}\frac{P_2}{P_1}\frac{s - s^+}{s^+} \frac{s_0 s_1}{(s - s_0)(s - s_1)} \left(\frac{1 - s^+}{s^+} \frac{s_0 s_1}{(1 - s_0)(1 - s_1)} - 1 \right)^{-1}. \quad \text{(III.44)}$$

We must now still let s_0, s_1 vary so as to maximize (III.44) while satisfying (III.42). This is a more difficult task than before, because (III.44) is not a monotonic function of s_0 and s_1. Nevertheless it can be shown that if there exists a maximum value of (III.44) subject to (III.42), then it is obtained only when (III.42) becomes an equality. This again has the consequence that $A_G(t)$ is a one-pole expression that satisfies the analogue of (III.40) as an equality, while the analogue of (III.39) remains a strict inequality. It is thus the same function that we found before when looking for a two-pole lower bound.

The fact that the same two-pole function has appeared both as an upper and as a lower bound should not dismay us. It is due to the assumption, not necessarily valid, that we made at the beginning; namely, that the bound for $A(s)$ is a two-pole function. The main result that we have shown here is that even when the bound is a two-pole function, we can still always restrict ourselves to considering only one-pole bounds either for $A(s)$ or for $A_G(t)$.

Turning now to a discussion of these one-pole bounds, we find the situation very much simplified. Assuming the form

$$\frac{A_0}{s - s_0} \quad \text{(III.45)}$$

for the (upper or lower) bound on $A(s)$, we first use (III.30) to eliminate A_0. We thus obtain the form

$$\frac{s^+ - s_0}{s - s_0} A(s^+) \quad \text{(III.46)}$$

for the bounding function, and the forms

$$\frac{s^+ - s_0}{s_0} A(s^+) \leq \frac{1}{3} \frac{p_2}{p_1} \qquad (III.47)$$

$$\frac{(2s_0 - 1)(s^+ - s_0)}{s_0(1 - s_0)} A(s^+) \leq \frac{1}{3} \frac{p_2}{p_1} \qquad (III.48)$$

for the inequalities (III.28) and (III.29). We must now vary s_0 to maximize (minimize) (III.46) in order to obtain an upper (lower) bound for $A(s)$.

Assuming, as before, that $s > 1$ and $s^+ < 0$, it is clear that an upper (lower) bound will be obtained by increasing (decreasing) s_0 as far as possible, subject to (III.47) and (III.48). The process of increasing s_0 will have to stop when (III.48) becomes an equality. Similarly, the process of decreasing s_0 will have to stop when (III.47) becomes an equality. In this way we obtain the following upper bound for $A(s)$

$$A(s) < \frac{s^+ - s_0}{s - s_0} A(s^+) \,, \qquad (III.49)$$

where s_0 is the solution of

$$\frac{(s^+ - s_0)(2s_0 - 1)}{s_0(s - s_0)} A(s^+) = \frac{1}{3} \frac{p_2}{p_1} \qquad (III.50)$$

that satisfies

$$\frac{2}{3} < s_0 < 1 \,. \qquad (III.51)$$

[Since (III.50) is a quadratic equation for s_0, it has two solutions. By using one of the Hashin-Shtrikman bounds which, for $A(s^+)$, reads

$$A(s^+) > \frac{2}{9} \frac{p_2}{p_1} \frac{1}{s^+ - \frac{2}{3}} \quad \text{for} \quad s^+ < 0 \,, \qquad (III.52)$$

we can show that one of these solutions indeed satisfies (III.51). The other solution is always negative, and hence unacceptable.] Similarly, the lower bound obtained in this way for $A(s)$ has the form

$$A(s) > \frac{s^+ - s_0}{s - s_0} A(s^+) \,, \qquad (III.53)$$

where s_0 is now the solution of

$$\frac{s^+ - s_0}{s_0} A(s^+) = \frac{1}{3} \frac{p_2}{p_1} \,. \qquad (III.54)$$

Using (III.52), we can show that s_0 satisfies

$$0 < s_0 < \frac{2}{3} \,. \qquad (III.55)$$

The two inequalities (III.52), (III.55) are required for consistency so that the second of the two inequalities (III.47), (III.48) (i.e., the one that is not an equality) is still obeyed as a (strict) inequality.

Translated into bounds for $F(s)$, the upper and lower bounds on $A(s)$ become upper and lower bounds for $F(s)$, respectively,

$$F(s) \leq F_g(s) \equiv \frac{p_1(s - s_0)}{(s - s_0)(s - \frac{1}{3} p_2) - \frac{1}{3} \frac{p_2 s_0(1 - s_0)}{2 s_0 - 1}} , \qquad (III.56)$$

with s_0 determined by (III.50), (III.51), and

$$F(s) \geq F_e(s) \equiv \frac{p_1(s - s_0)}{s(s - s_0 - \frac{1}{3} p_2)} , \qquad (III.57)$$

with s_0 determined by (III.54).

We must also consider the one-pole bounds on $A_G(t)$. The analogue of (III.53), (III.54) leads again to the bound $F_e(s)$ for $F(s)$. The analogue of (III.49)-(III.51) leads, however, to a two-pole bound for $A(s)$, and finally to the following three-pole bound for $F(s)$

$$F(s) \leq F_h(s) \equiv \frac{1}{s} \frac{p_1(s - s_0)(s - \frac{2}{3}) - \frac{1}{3} \frac{p_1 s_0(1 - s_0)}{1 - 2 s_0}}{(s - s_0)(s - 1 + \frac{1}{3} p_1) - \frac{1}{3} \frac{p_1 s_0(1 - s_0)}{1 - 2 s_0}} , \qquad (III.58)$$

where s_0 is the solution of

$$\frac{(s^+ - s_0)(2 s_0 - 1)}{s_0(1 - s_0)} A_G(1 - s^+) = \frac{1}{3} \frac{p_1}{p_2} \qquad (III.59)$$

that also satisfies

$$0 < s_0 < \frac{1}{3} . \qquad (III.60)$$

[As we mentioned in the bracketed remarks related to (III.51) and (III.52), we can show that one of the solutions of (III.59) obeys (III.60), while the other is a negative number.] We note that the function $A_h(s)$ related to $F_h(s)$ is a two-pole upper bound on $A(s)$, and also that it is the same function that appeared earlier in this section as a candidate for both upper and lower bounds. From the present discussion, it is clear that $F_h(s)$ is now only a candidate for the role of upper bound on $F(s)$. The actual upper bound is in fact the greater of the two numbers $F_g(s)$ and $F_h(s)$.

It can be shown that in general (i.e., not restricting ourselves to the case $s > 1, s^+ < 0$), F_e always gives one of the bounds on F, while the other bound is given by either F_g or F_h, whichever lies further away from F_e.

Comparing these bounds with the bounds of (III.19) for ϵ_e, we find that one of the latter bounds is identical with F_e. The other bound of (III.19) is, however, different (i.e., less stringent) than either F_g or F_h. In fact, it is equivalent to

$$F(s) \leq F_f(s) \equiv \frac{p_1(s - s_0)}{(s - s_0)(s - \frac{1}{3}p_2) - \frac{2}{3}p_2(1 - s_0)} \quad , \tag{III.61}$$

where s_0 is now obtained by treating (III.25) as an equality. Close examination shows that $F_f(s)$ violates (III.27), and that therefore it is an overly stringent bound.

As an example we now compute some of the bounds on ϵ_e in the following case

$$p_1 = p_2 = \frac{1}{2}, \quad \epsilon_1 = \epsilon_1^+ = 1, \quad \epsilon_2 = 10, \quad \epsilon_2^+ = \frac{1}{3}, \quad \epsilon_e^+ = \frac{3}{5},$$

$$s = \frac{10}{9}, \quad s^+ = -\frac{1}{2}, \quad F(s^+) = -\frac{4}{5} \quad . \tag{III.62}$$

Listed in the table below are the Wiener bounds (W), the Hashin-Shtrikman bounds (HS), the bounds of (III.19) (B_1), and the bounds of (III.57), (III.56), and (III.58) (B_2). An upper (lower) bound is denoted by ϵ_u (ϵ_ℓ), and the last two bounds, both of them lower bounds, are distinguished by an additional suffix g or h to denote whether they arose from F_g or F_h. The difference between the actual upper and lower bounds in each category is denoted by $\Delta\epsilon$.

W			HS			B_1			B_2			
ϵ_u	ϵ_ℓ	$\Delta\epsilon$	ϵ_u	ϵ_ℓ	$\Delta\epsilon$	ϵ_u	ϵ_ℓ	$\Delta\epsilon$	ϵ_u	$\epsilon_{\ell g}$	$\epsilon_{\ell h}$	$\Delta\epsilon$
5.5000	1.8182	3.68	4.6712	2.8832	1.79	4.5357	3.2836	1.25	4.5357	3.6452	3.4367	1.20

Clearly, the bound $\epsilon_{\ell h}$ (B_2) constitutes an improvement over the bound ϵ_ℓ (B_1).

Rigorous bounds can also be constructed when one or both of ϵ_1, ϵ_2 are complex, and hence ϵ_e is in general complex too. In that case, one can proceed by separating (II.12a) into its real and imaginary parts, treating $\text{Im} F(s)$ (for example) as given, and looking for a maximum or minimum value of $\text{Re} F(s)$. The equation for $\text{Im} F(s)$ is thus treated as an additional constraint. Such a program has been carried out in detail (Bergman 1980, 1981 a,b; Milton 1980, 1981), and the results are always in the form of a limited region in the complex F-plane where the allowed values of $F(s)$ must lie. It should be noted, of course, that complex values of ϵ are usually found only by going to finite frequencies. In that case, the electrostatic theory used here becomes an approximation which is only valid as long as both the wavelength and any possible skin depth are large compared to any scale of the microscopic inhomogeneities in the composite. The case where this requirement is violated is discussed in Section VI.

IV. CALCULATION OF THE ELECTROSTATIC RESONANCES OF A TWO-COMPONENT COMPOSITE

It is often possible to describe a two-component composite as being made of a host material ϵ_2, in which are embedded separate, non-overlapping grains of ϵ_1 material. In that case, the function $\theta_1(\vec{r})$ can be written as a sum of individual grain θ-functions

$$\theta_1(\vec{r}) = \sum_a \theta_a(\vec{r}) \qquad (IV.1)$$

where a is a grain index, and θ_a is defined by

$$\theta_a(\vec{r}) = \begin{cases} 1 & \text{for } \vec{r} \text{ inside the grain } a \\ 0 & \text{elsewhere} \end{cases} \qquad (IV.2)$$

Consequently, the operator Γ can also be written as a sum of individual grain operators

$$\Gamma = \sum_a \Gamma_a$$

$$\Gamma_a \phi \equiv \int dV' \theta_a(\vec{r}') \frac{\partial G(\vec{r},\vec{r}')}{\partial \vec{r}'} \cdot \frac{\partial \phi(\vec{r}')}{\partial \vec{r}'} \quad . \qquad (IV.3)$$

It is usually easier to find the eigenstates of each individual Γ_a than to find those of Γ. In that case, it is a good strategy to try to expand any solution of

$$\phi = z + u \Gamma \phi \qquad (IV.4)$$

or of the corresponding eigenvalue equation

$$s\phi = \Gamma\phi \qquad (IV.5)$$

in terms of these individual grain eigenfunctions, defined by

$$s_{a\alpha} \phi_{a\alpha} = \Gamma_a \phi_{a\alpha} \quad . \qquad (IV.6)$$

Such an expansion would have the form

$$\theta_1^+(\vec{r}) \phi(\vec{r}) = \sum_{a\alpha} A_{a\alpha} \theta_a^+(\vec{r}) \phi_{a\alpha}(\vec{r}) \quad , \qquad (IV.7)$$

where $A_{a\alpha}$ are the expansion coefficients. The θ_1-function must appear on the ℓ.h.s. because we can only hope to expand $\phi(\vec{r})$ in this way at points within ϵ_1 material. Since the scalar product of (II.4) involves integration only over the ϵ_1 volume, the completeness property of eigenstates of Γ applies only within that volume. Similarly, the functions $\phi_{a\alpha}$ for a fixed grain a, though defined everywhere, constitute a complete orthonormal set only within the volume of that grain. That is why on the r.h.s. of (IV.7) we have to restrict each of these functions to the volume of its particular grain by including θ_a. Finally, the (+) superscript attached to the θ-functions in (IV.7) means that we are to take them to equal 1 over a volume that is infinitesimally larger than the usual volume. This is done in order to prevent spurious δ-functions from appearing as a result of differentiating these θ-functions at the ϵ_1, ϵ_2 interface. The need for defining these θ^+ functions is also one of the reasons why we assumed that the different grains were non-overlapping. (It now appears that it is possible to extend this formalism to a system of overlapping grains - the price includes having to deal explicitly with these δ-functions, as well as with the fact that eigenstates of different individual grain operators are not orthogonal when the grains overlap - see Kantor and Bergman 1981 c.)

Substituting the expansion (IV.7) into (IV.4), and using the usual methods, we get a system of linear equations for the expansion coefficients

$$A_{a\alpha} = z_{a\alpha} + u \sum_{b\beta} \Gamma_{a\alpha,b\beta} A_{b\beta} \quad . \tag{IV.8}$$

Here $z_{a\alpha}$ are the expansion coefficients obtained when $\theta^+ z$ is expanded in a series of $\theta_a^+ \phi_{a\alpha}$, while the coefficients appearing in the sum are matrix elements of Γ between two individual grain eigenfunctions

$$\Gamma_{a\alpha,b\beta} \equiv \langle \theta_a^+ \phi_{a\alpha} | \Gamma | \theta_b^+ \phi_{b\beta} \rangle$$

$$= \int dV \int dV' \theta_a(\vec{r}) \theta_b(\vec{r}\,') \frac{\partial \phi_{a\alpha}^*(\vec{r})}{\partial \vec{r}} \cdot \frac{\partial^2 G(\vec{r},\vec{r}\,')}{\partial \vec{r} \, \partial \vec{r}\,'} \cdot \frac{\partial \phi_{b\beta}(\vec{r}\,')}{\partial \vec{r}\,'}$$

$$= s_{b\beta} \int dV \, \theta_a(\vec{r}) \, \vec{\nabla} \phi_{a\alpha}^* \cdot \vec{\nabla} \phi_{b\beta} \quad . \tag{IV.9}$$

The last expression for the matrix element is especially convenient for practical calculations - it has the form essentially of an overlap integral of the product $\vec{\nabla} \phi_{a\alpha}^* \cdot \vec{\nabla} \phi_{b\beta}$ integrated over the volume of just <u>one</u> of the two grains (i.e., the grain a).

This discussion is reminiscent of the so called "tight binding approach" to the calculation of electronic states (i.e., solutions of Schrödinger's equation) in a collection of many atoms. There too, one starts from the (electronic) eigenstates of individual atoms, and then uses them as a basis in which to expand the solution of the many-atoms problem.

As in the electronic problem, here too, a significant simplification occurs if the ε_1 grains (analogous to individual atoms) are all identical and arranged in a periodic array. In that case, if we use a vector index $\vec{a}$ to denote the position of an individual grain, then, by a proper choice of the normalization constant, the functions $\phi_{\vec{a}\alpha}(\vec{r})$ for different grains can be made to depend only on $\vec{r} - \vec{a}$

$$\phi_{\vec{a}\alpha}(\vec{r}) = \phi_\alpha(\vec{r} - \vec{a}) \quad . \tag{IV.10}$$

Consequently, the $z_{\vec{a}\alpha}$ coefficients are independent of $\vec{a}$, and the $A_{\vec{a}\alpha}$ coefficients can be shown to obey a Bloch-theorem, i.e., they always have the form

$$A_{\vec{a}\alpha} = A_\alpha(\vec{k}) \, e^{i\vec{k}\cdot\vec{a}} \quad . \tag{IV.11}$$

Equation (IV.8) is then replaced by

$$A_\alpha(\vec{k}) = z_\alpha \, \delta_{\vec{k},\vec{0}} + u \sum_\beta \Gamma_{\alpha\beta}(\vec{k}) \, A_\beta(\vec{k}) \tag{IV.12a}$$

$$\Gamma_{\alpha\beta}(\vec{k}) \equiv \sum_{\vec{a}-\vec{b}} \Gamma_{\vec{a}\alpha,\vec{b}\beta}\, e^{-i\vec{k}\cdot(\vec{a}-\vec{b})} \quad . \tag{IV.12b}$$

If we omit $z_{a\alpha}$ or z_α in (IV.8) or (IV.12a), respectively, then these become equations for the expansion coefficients of the eigenfunctions of Γ. These are just the eigenvectors of the matrix in each of these equations. With the help of these eigenvectors, we can rewrite (II.12a) in the form

$$F(s) = \frac{u}{V} \sum_{a\alpha} z^*_{a\alpha} A_{a\alpha} \tag{IV.13}$$

for a general system of non-overlapping grains, or in the form

$$F(s) = \frac{uN}{V} \sum_\alpha z^*_\alpha A_\alpha(\vec{k}=0) \tag{IV.14}$$

for a periodic array of identical grains. The factor N is the total number of unit cells in the system. Note that only the $\vec{k} = 0$ eigenvectors are needed to determine ϵ_e - the nonzero $\vec{k}$ eigenvalues have zero weight in the pole spectrum of F(s). That is why the pole spectrum remains discrete in this case, even though the eigenvalue spectrum is continuous (for systems with an infinite volume).

The procedure described in this section is especially suitable when the individual grains have very simple and highly symmetrical shapes. E.g., when the ϵ_1 grains are all spherical, the relevant individual grain eigenstates and eigenvalues can all be obtained in closed analytical form, and the same is true of the two-body overlap integrals. Thus, the matrix elements $\Gamma_{a\alpha,b\beta}$ can be set up analytically, and it is only when one wants to solve (IV.8) or (IV.12) that a numerical computation must be resorted to. This has actually been done in detail for a simple cubic array of identical spherical inclusions, with very good results (Bergman 1979 a), and also for a quasi-random system of spheres, where the results could be compared with experiments (Kantor and Bergman 1981 b).

In summary, the approach described in this section can take maximum advantage of any symmetries or simplifying features of individual grains, even when the entire composite has a much lower symmetry. We might also add that McPhedran and Milton (1981) have developed an alternative way of using the matrix $\Gamma_{a\alpha,b\beta}$ for obtaining ϵ_e. In their approach, the matrix is first used to calculate the higher moments of the pole spectrum s_n, defined in (II.15). These higher moments are then used to generate a series of successively better upper and lower bounds on ϵ_e. This has certain advantages over a direct calculation of ϵ_e, since it is unnecessary either to invert the matrix or diagonalize it. At the same time, it provides an estimate of the error in the calculation.

V. ELECTROSTATIC PROBLEMS IN MULTICOMPONENT COMPOSITES

When more than two components are involved, the bulk effective dielectric constant must be discussed in terms of an analytic function of more than one variable

$$F(s^{(1)}\ldots s^{(m)}) \equiv 1 - \frac{\epsilon_e(\epsilon_1\ldots\epsilon_{m+1})}{\epsilon_{m+1}} = \sum_{i=1}^{m}\frac{u_i}{V}\int dV\,\theta_i\,\frac{\partial\phi}{\partial z} \quad . \tag{V.1}$$

In general, the singularities of this function lie on $(m-1)$-dimensional hypersurfaces in the complex space of $s^{(1)}\ldots s^{(m)}$. For a system of non-overlapping grains with an exactly known microgeometry, the method of the previous section can obviously be generalized so as to determine these hypersurfaces: Fixing the values of, say, $s^{(2)}\ldots s^{(m)}$, we can find the resonance values of $s^{(1)}$ as eigenvalues of (II.1). The eigenvectors, which, like the eigenvalues, now depend on $s^{(2)}\ldots s^{(m)}$, can be used to construct the weights in a pole expansion of F in the variable $s^{(1)}$.

The problem of deriving exact bounds for $F(s^{(1)}\ldots s^{(m)})$ when the microgeometry is not precisely known is best handled by selecting an appropriate trajectory $s^{(1)}(s)\ldots s^{(m)}(s)$. If this is done properly, the resulting function of one variable

$$F(s) \equiv F(s^{(1)}(s)\ldots s^{(m)}(s)) \tag{V.2}$$

can be shown to have analytic properties similar to those that are found in the two-component problem. The methods of Section III can then be applied to this function in order to derive rigorous bounds when various types of partial information are available about the composite (see Bergman 1978 a,b).

VI. EXTENSION OF THE APPROACH TO OTHER THAN ELECTROSTATIC PROBLEMS

The basic idea - focusing on the resonances of the problem and using them in order to expand the real problem - is applicable to other bulk properties of composites, and also to scattering problems.

The problem of calculating the bulk effective elastic stiffness tensor $C_{ijk\ell}^{(e)}$ can be set up in a form that is similar to (II.1) and (II.10), namely (we assume a two-component composite),

$$\epsilon_{ij}(\vec{r}) = \epsilon_{ij}^o + \sum_{k\ell mn}\int dV'\,\theta_1(\vec{r}')\,G_{ijk\ell}(\vec{r},\vec{r}')\,\delta C_{k\ell mn}\,\epsilon_{mn}(\vec{r}')$$

$$\epsilon^o C^{(e)}\epsilon^o \equiv \frac{1}{V}\int dV\,\epsilon^o\,C(\vec{r})\,\epsilon(\vec{r})$$

$$C_{ijk\ell}(\vec{r}) = C_{ijk\ell}^{(1)}\,\theta_1(\vec{r}) + C_{ijk\ell}^{(2)}\,\theta_2(\vec{r})$$

$$= C^{(2)} + \theta_1(\vec{r})\,\delta C$$

$$\delta C \equiv C^{(1)} - C^{(2)} \tag{VI.1}$$

Here $\epsilon_{ij}(\vec{r})$ is the local strain tensor in the composite, while ϵ_{ij}^o is the (constant) strain tensor that would exist if the same boundary conditions were applied to a homogeneous solid. The C's are elastic stiffness tensors, and wherever it is convenient we have omitted the tensor indices and allowed the summations over them to be implied.

Similarly to the case of the multicomponent composite dielectric, here too, the problem depends upon more than one (complex) variable, since even in the case of an isotropic material there are two elastic constants for each component. By applying methods similar to those described in Section V for the multicomponent dielectric composites, a representation of $C^{(e)}$ in terms of elastostatic resonances can be obtained. This has been used to calculate $C^{(e)}$ for ordered two-dimensional arrays of cylindrical fibers (Kantor and Bergman 1981 a), and to derive some new exact bounds for systems with an unknown microgeometry (Kantor and Bergman 1981 c).

A different type of problem is the scattering of electromagnetic waves by a macroscopic scatterer. By that we mean that a monochromatic plane wave which is propagating through a homogeneous medium characterized by ε_2 (we ignore the magnetic permeability, i.e., assume that $\mu = 1$ everywhere), is scattered because a (limited) part of space is occupied by a different material - ε_1. When the characteristic dimensions of the scatterer are not small compared to the wavelength or the skin depth, the static approximation is inapplicable. The problem can be cast in the form of an integral equation for the electric field $\vec{E}$

$$\vec{E}(\vec{r}) = \vec{E}_0(\vec{r}) + u \int dV' \theta_1(\vec{r}') \; \vec{\vec{G}}(\vec{r} - \vec{r}', k_2) \cdot \vec{E}(\vec{r}')$$

$$\equiv E_0 + u \, \Gamma \, E \tag{VI.2}$$

where $\vec{E}_0(\vec{r})$ is the incident plane wave, $k_2 \equiv \sqrt{\varepsilon_2}\, \omega/c$ is the wave number in the ε_2 medium, $u \equiv 1/s \equiv 1 - \varepsilon_1/\varepsilon_2$, and where the last line of (VI.2) serves to define the linear operator Γ. The symbol $\vec{\vec{G}}$ stands for a tensor Green's function that satisfies boundary conditions of an outgoing spherical wave at infinity

$$G_{\alpha\beta}(\vec{R}, k) \equiv (k^2 \delta_{\alpha\beta} + \nabla_\alpha \nabla_\beta) \frac{e^{ikR}}{-4\pi R} \quad . \tag{VI.3}$$

The operator Γ, while symmetric, is non-hermitian when one introduces the analogue of (II.4) as the definition of a scalar product

$$(E_1, E_2) \equiv \int dV \; \theta_1(\vec{r}) \; \vec{E}_1^*(\vec{r}) \cdot \vec{E}_2(\vec{r}) \quad . \tag{VI.4}$$

Nevertheless, we can still find the eigenfunctions (in this case they are vector fields) and eigenvalues (they are complex) of Γ, which now obey the following bi-orthogonality relations

$$\Gamma |t\rangle = t|t\rangle$$

$$\langle t^*|\Gamma = \langle t^*|t$$

$$\langle t^*|s\rangle = 0 \quad \text{for different states} \tag{VI.5}$$

If we can extend the bi-orthogonality to include also the normalization property

$$\langle t^*|t\rangle = 1 \quad , \tag{VI.6}$$

something that is not automatically ensured but must be verified in each case, then the set of eigenstates can be shown to be complete. We can then expand Γ as

$$\Gamma = \sum_t |t\rangle t \langle t^*| \quad , \tag{VI.7}$$

and use this to write the <u>scattered</u> part of the field E_{sc} in terms of the states $|t\rangle$

$$E = \frac{1}{1 - u\Gamma} E_0 = E_0 + \frac{\Gamma}{s - \Gamma} E_0 \equiv E_0 + E_{sc} \tag{VI.8}$$

$$E_{sc} = \sum_t |t\rangle \frac{t}{s - t} \langle t^*|E_0\rangle \quad . \tag{VI.9}$$

It is important to realize that although the states $|t\rangle$ only form a complete set within the volume of the scatterer ϵ_1, the expansion of (VI.9) is valid everywhere. This is so because the eigenfunctions $|t\rangle$ obey the same differential equation as the total field E. Another way of perceiving this is to note that (VI.7) represents the operator Γ everywhere. Thanks to this fact, the expansion (VI.9) can be used in the asymptotic region, where it gives the asymptotic form of E_{sc} in terms of the asymptotic forms of $|t\rangle$. Note however that we need the form of $|t\rangle$ inside the scatterer in order to calculate the partial scattering amplitude into that state, i.e.,

$$\frac{t}{s - t} \langle t^*|E_0\rangle \quad . \tag{VI.10}$$

This approach has been carried out in detail for the case of a scatterer in the form of a collection of spheres (Bergman and Stroud 1980). In that case, it is worthwhile to use the eigenfunctions of individual spheres as a basis set in Hilbert space. All the multiple scattering effects are then automatically taken into account when the matrix of Γ in that basis is diagonalized. As in the electrostatic problem, the off-diagonal matrix elements are expressible in terms of overlap intergrals between eigenstates of different spheres. Consequently, once one has determined the eigenstates and the two-body overlap integrals, the problem of scattering by three or more spheres is not significantly more difficult than the two-sphere problem.

Finally, we would like to mention the fact that the quantum scattering problem can also be cast in a similar form. Starting from the Lippmann-Schwinger integral equation for scattering by a scalar potential $V(\vec{r})$

$$\psi(\vec{r}) = e^{i\vec{k}\cdot\vec{r}} - \frac{m}{2\pi\hbar^2} \int dV' \frac{e^{ik|r - r'|}}{|r - r'|} V(\vec{r}')\psi(\vec{r}') \quad , \tag{VI.11}$$

we first multiply it by $\sqrt{V(\vec{r})}$ and rewrite it as the last of the following equations

$$\rho(\vec{r}) \equiv \sqrt{V(\vec{r})}\, \psi(\vec{r}) \tag{VI.12a}$$

$$\Gamma\rho \equiv -\frac{m}{2\pi\hbar^2} \int dV' \sqrt{V(\vec{r})} \frac{e^{ik|r-r'|}}{|r-r'|} \sqrt{V(\vec{r}')} \rho(\vec{r}')$$

$$\rho = \rho_o + \Gamma\rho \quad . \tag{VI.12b}$$

If $V(\vec{r}) \neq 0$ only in a limited region of space, which is often the case to a very good approximation, then we can characterize that region by an appropriate θ-function and define a scalar product that includes integration only over that limited region

$$(\psi_1, \psi_2) \equiv \int dV \; \theta(\vec{r}) \psi_1^*(\vec{r}) \psi_2(\vec{r}) \quad . \tag{VI.13}$$

We can then proceed, as in the e.m. scattering problem, to calculate the eigenstates of the operator Γ and use them as a basis set in which to expand the actual scattering problem. Because Γ is again a non-hermitian operator, bi-orthogonality relations will have to be used in manipulating the basis set. It is expected that this method will be especially useful in treating problems where multiple scattering is important, and where the symmetry of individual elements of the scatterer is high while the total scatterer has a lower symmetry.

VII. CONDUCTIVITY OR CONNECTIVITY THRESHOLDS

Composite materials that are made of a metal-insulator mixture can either be conducting or insulating, depending upon the microgeometry. Often, by changing a single microstructural parameter, e.g., the overall concentration of metal, it is possible to switch from one type of behavior to the other. One very simple (but not very practical) system where this occurs is a periodic array of conducting spheres in an insulating background. As the spheres are increased in size, such a transition will occur at the point where they just begin to touch each other. The transition in the case of a simple cubic array of spheres has been studied in some detail (Bergman 1979a).

A somewhat different case is where the metal-insulator system has a random microstructure. It should then really be represented by an ensemble of systems with different detailed microgeometries, and the bulk effective conductivity σ_e (or any other property, for that matter) should be averaged over all members of the ensemble. The conductivity threshold is then associated with a percolation threshold of the metallic component.

In general, a conductivity threshold in a metal-insulator composite is always connected with a geometric connectivity threshold for the metallic component.

Since the function $F(s)$ of (II.12) represents the bulk d.c. conductivity σ_e of a two-component composite as effectively as it does ε_e, we now focus attention upon it, recalling that the variables u, s should now be related to the pure component conductivities σ_1, σ_2

$$u \equiv \frac{1}{s} \equiv 1 - \frac{\sigma_1}{\sigma_2} \quad , \tag{VII.1}$$

while

$$F(s) \equiv 1 - \frac{\sigma_e}{\sigma_2} \quad . \tag{VII.2}$$

Clearly, if σ_2 is the metal and σ_1 is the insulator, then $s = 1$ and we have

$$F(1) = 1 \quad \text{if} \quad \sigma_e = 0$$

$$0 < F(1) < 1 \quad \text{if} \quad \sigma_e > 0 \quad . \tag{VII.3}$$

On the other hand, if σ_1 is the metal and σ_2 is the insulator, then we should be looking at $s = 0$. It is then more convenient to think of a material with the same microgeometry, but where $\sigma_1 = \infty$ while σ_2 is finite. This again leads to $s = 0$, and it is now clear that

$$-\infty < F(0) < 0 \quad \text{if} \quad \sigma_e < \infty$$

$$F(0) = -\infty \quad \text{if} \quad \sigma_e = \infty \quad . \tag{VII.4}$$

To be quite specific, the last line means that $s = 0$ is a pole of $F(s)$ with a finite weight. Thus, the connectivity properties of $\sigma_1 (\sigma_2)$ are directly related to the value of $F(0)$ ($F(1)$).

In a random composite, we are really interested in the average of $F(s)$ over the ensemble. Thus, the spectrum of poles, even if it is discrete for a specific system, gets smeared into a continuous spectrum by the averaging process. Consequently $F(s)$ now has a branch cut on the real axis somewhere between $s = 0$ and $s = 1$.

Quite often, the system or the statistical ensemble is characterized by a parameter, such as the volume fraction p_1 or $p_2 = 1 - p_1$, that can be changed continuously until at some particular value p_c the system switches from conducting to nonconducting. We will now discuss the behavior of $F(s)$ near such a conductivity or connectivity threshold.

From the definitions of $F(s)$ and $E(s)$, (II.12a) and (III.17) respectively, we can deduce the following relationship

$$(1 - F(s))(1 - E(s)) = 1 - \frac{1}{s} \tag{VII.6}$$

that holds when E and F correspond to any specified microgeometry. Since we want to use (VII.6) also for the case of a random system, where E and F correspond to ensemble averages over different microgeometries, some words of caution are in order.

Although in general, the average of E·F is not equal to the product of averages of E and of F, we can argue that equality should always hold if the ensemble is to have any physical significance. Unless the distribution of values of σ_e for systems in the ensemble is sufficiently narrow so that calculating $\langle \sigma_e \rangle$ yields the same result as calculating $\langle 1/\sigma_e \rangle^{-1}$, both values are patently useless. After all, an experiment is always performed with a single sample, and the ensemble is only a mathematical tool for representing the degree of our uncertainty about the actual microgeometry of that sample. An ensemble that yields a wide range of values for σ_e simply means that the set of samples is not well enough characterized, i.e., that its relevant properties (σ_e in this case) are irreproducible from one sample to the next.

We will now consider the conductivity or connectivity threshold of σ_2, hence we wish to take $\sigma_2 > 0$, $\sigma_1 = 0$, i.e., s = 1. At that point, F(1) and E(1) exhibit the following properties

$$E(1) = 1, \quad 0 < F(1) < 1 \quad \text{if} \quad p_2 > p_c$$

$$F(1) = 1, \quad 0 < E(1) < 1 \quad \text{if} \quad p_2 < p_c \quad . \tag{VII.7}$$

Note that the inequalities in this equation are strict inequalities - the possibility of an equality is not allowed.

From (VII.6) it is now clear that when $p_2 < p_c$, then $1 - F(s) \sim 1 - s$ for $s \to 1$. Consequently, the derivative $F'(s)$ remains finite as $s \to 1$. As for the other derivative $E'(s)$, as $s \to 1$ it must either remain finite, or at most diverge more weakly than $1/(1-s)$ (otherwise E(1) would not be finite). Taking the derivative of (VII.6) and using these results to set s = 1, we now get

$$F'(1)(1 - E(1)) = -1 \quad . \tag{VII.8}$$

If we now let $p_2 \to p_c^-$, E(1) must tend to 1, and consequently, $F'(1) \to \infty$.

Viewed as a function of s for $p_2 = p_c$, $F'(s)$ must diverge as $s \to 1$, but less weakly than $1/(1-s)$ so that F(1) remains finite. This means that s = 1 must be a singularity of F(s). Consequently, in the random percolative case, the branch cut in F(s) must extend all the way to s = 1, which is a branch point, at $p_2 = p_c$. From (VII.6) we can now deduce that, for $p_2 = p_c$, $E'(s)$ must also diverge as $s \to 1$, and that the product $E'(s)F'(s)$ diverges as $1/(1-s)$.

I would have liked to argue that away from p_c, i.e., for $p_2 < p_c$, both E(s) and F(s) are analytic at s = 1. However, I cannot prove this assertion in general. In fact, I suspect that for particular ensembles it may not be true. However, it is clear that for a particular finite (or infinite periodic) system, with a well defined and smooth microgeometry where σ_2 is unconnected or non-percolating, the highest singularity of F(s) is a pole of finite weight that appears a finite distance below s = 1. It is then plausible that even if an ensemble of such systems or an infinite but random system is considered, the highest singularity will still lie below s = 1. (A more detailed physical argument for making this plausible is given in Bergman 1978 b).

Turning to the regime above the threshold for connectivity of the σ_2 component, namely $p_2 > p_c$, we can proceed in complete analogy with the previous discussion except for switching the roles of E(s) and F(s). In this way we can show that $E'(s)$ must remain finite at s = 1, while $F'(s)$ either remains finite as $s \to 1$, or at most diverges more weakly than $1/(1-s)$. As before, physical plausibility arguments can be made to suggest that $s = 1$ is in fact a regular point of both $E(s)$ and $F(s)$ for $p_2 > p_c$ (see Bergman 1978 b).

It is worthwhile to point out that this type of predicted behavior is in complete agreement with the effective medium approximation, which is the only solvable model that exhibits a percolation-type conductivity threshold (see Bergman 1978 b). It is also in agreement with asymptotic analyses of series calculations of $F(s)$ in powers of $u = 1/s$ for a random-resistor-network (Bergman and Kantor 1981).

The fact that a conductivity threshold is invariably connected with the singularity of $F(s)$ at $s = 1$ (or at $s = 0$, depending upon whether σ_2 or σ_1 is the metallic component), as we have shown above, turns out to be a very useful tool in obtaining some important general relationships. It has thus been possible to predict that in a metal-insulator composite, the real part of the dielectric constant diverges at the conductivity threshold, as well as to relate the critical exponents for the singular behavior of the dielectric constant and the conductivity (Bergman and Imry 1977). It has also been possible to make predictions regarding the frequency dependence of the above mentioned polarization catastrophe at a metal-insulator percolative transition as well as at a normal-superconducting percolative transition (Stroud and Bergman 1981).

VIII. ACKNOWLEDGEMENTS

It is a pleasure to acknowledge useful conversations and correspondence with Graeme Milton of Cornell University and The University of Sydney. It is also a pleasure to thank Phyllis Dolan of The Ohio State University for her great devotion in getting this manuscript typed in a camera-ready form that is so pleasant to look at and read.

BIBLIOGRAPHY

Bergman, D. J. 1978a AIP Conf. Proc. No. 40, pp. 46-61.
Bergman, D. J. 1978b Physics Reports 43 377-407.
Bergman, D. J. 1979a J. Phys. C 12 4947-4960.
Bergman, D. J. 1979b Phys. Rev. B 19 2359-2368.
Bergman, D. J. 1980 Phys. Rev. Lett. 44 1285-1287.
Bergman, D. J. 1981a Phys. Rev. B 23 3058.
Bergman, D. J. 1981b, To appear in Annals of Physics.
Bergman, D. J. 1981c, Unpublished.
Bergman, D. J. and Imry, Y. 1977 Phys. Rev. Lett. 39 1222-1225.
Bergman, D. J. and Kantor, Y. 1981 J. Phys. C, to be published.
Bergman, D. J. and Stroud, D. 1980 Phys. Rev. B 22 3527.
Dantzig, G. B. 1963 Linear Programming and Extensions, Princeton Univ. Press, Princeton, New Jersey.
Hashin, Z. and Shtrikman, S. 1962 J. Appl. Phys. 33 3125-3131.
Kantor, Y. and Bergman, D. J. 1981a, Submitted to J. Mech. Phys. Sol.
Kantor, Y. and Bergman, D. J. 1981b, To appear in J. Phys. C: Solid State Physics.
Kantor, Y. and Bergman, D. J. 1981c, In preparation.
McPhedran, R. C. and Milton, G. 1981, To appear in Applied Physics.
Milton, G. 1980 Appl. Phys. Lett. 37 300-302.

Milton, G. 1981, To appear in J. Appl. Phys.
Prager, S. 1969 J. Chem. Phys. 50 4305-4312.
Schulgasser, K. 1976 J. Math. Phys. 17 378-381.
Stroud, D. and Bergman, D. J. 1981, In preparation.
Weiner, O. 1912 Abh. Math. Phys. Kl. Sächs. Akad. Wiss. Leipz. 32 509.

ELASTIC WAVES IN FLUID-SATURATED POROUS MEDIA

James G. Berryman
Bell Laboratories
Whippany, N.J.   07981/USA

1. Introduction

Porous materials arise in a variety of geophysical contexts and engineering applications. Typical examples of fluid-saturated porous materials include (1) porous, air-filled, sound-absorbing tiles used in noise control and (2) water- or oil-saturated sedimentary rocks studied by earthquake seismologists and oil-exploration geophysicists.

A general theory of elastic wave propagation in fluid-saturated porous media was developed by Biot [1-3]. Although Biot's theory was published 25 years ago, it has only been within the last few years that the slow compressional wave predicted by Biot's theory has been observed in unambiguous experiments [4-6]. Further, only recently has it been realized that *a priori* estimates of several of the phenomenological parameters in Biot's equations were available [5,7-13]. In the past, Biot's equations were used primarily as a phenomenological tool, using curve fitting methods to obtain estimates of those parameters which were difficult to measure or estimate directly [6,14-15]. Now the scientific basis for Biot's equations is more firmly established and many of the uncertainties involved in applying Biot's equations to practical problems have been resolved.

In this paper, we will review the recent progress in understanding Biot's theory, present the model equations found to be useful in applications, and finally give a brief comparison of the theory's predictions to some recent experimental data.

2. Model equations

A fluid-saturated porous material is characterized by the presence of a porous frame and a fluid which fills the interconnected pore space. For example, the porous frame might consist (1) of an open-cell rubber foam or (2) of sand grains cemented together or held together at some depth by over-burden pressure. The interconnected pore space occupies the fraction β of the total volume known as the porosity. Any part of the fluid which is trapped in pores detached from the interconnected pore space is said to occupy the "secondary porosity." Such pockets of isolated fluid are treated as fluid inclusions modifying the elastic properties of the composite porous frame. For simplicity, we generally assume the secondary porosity vanishes but this assumption is not essential to the subsequent analysis.

To establish notation: The bulk modulus and density of the fluid are K_f and ρ_f, respectively. The bulk and shear moduli of the (dry) porous frame are K^* and μ^*. If the frame is composed of a single type of solid constituent grain (as we assume here), the bulk and shear moduli and density of the grains are K_g, μ_g, and ρ_g. The average displacement vector for the solid frame is $\vec{u}$. The fluid-displacement field $\vec{u}_f$ derived from the average volume flow of the fluid through the pores. The average fluid displacement relative to the frame is $\vec{w} = \beta(\vec{u}_f - \vec{u})$. For small strains, the frame dilatation is

2.1 $\qquad e = \vec{\nabla} \cdot \vec{u}$

The increment of fluid content ζ [3] is defined [16] as the change in pore space per unit volume minus the change in fluid volume per unit volume, i.e.,

2.2 $\qquad d\zeta = \beta(de - de_f) + d\beta.$

In (2.2), e_f is the average of the microscopic fluid dilatation which must be distinguished from

2.3 $\qquad \varepsilon = \vec{\nabla} \cdot \vec{u}_f.$

Since $\vec{u}_f$ is a quantity derived from the average flow of the fluid, ε includes a term due to flow in addition to the average microscopic dilatation e_f. Finally, the increment of fluid content is related to the relative displacement vector by

2.4 $\qquad \zeta = \vec{\nabla} \cdot \vec{w} = \beta(e - \varepsilon).$

We emphasize that $e_f \neq \varepsilon$ in general.

With these definitions for a linear medium isotropic on the macroscopic scale, Biot [3] shows that, if a strain energy functional E exists, it is a quadratic function of the strain invariants [17] $I_1 = e$, I_2 and of ζ having the form

2.5 $$2E = He^2 - 2Ce\zeta + M\zeta^2 - 4\mu^* I_2,$$

where

2.6 $$I_2 = e_x e_y + e_y e_z + e_z e_x - \frac{1}{4}(\gamma_x^2 + \gamma_y^2 + \gamma_z^2).$$

The normal strain components are e_x, e_y, e_z while the shear strain components are γ_x, γ_y, γ_z. Using *gedanken* experiments to analyze the effects of applying an isotropic hydrostatic pressure to the saturated medium [15,18,19], the coefficients H, C, and M have been shown to be given by

2.7 $$K \equiv H - \frac{4}{3}\mu^* = K^* + \sigma C,$$

2.8 $$C = \left[\left(\frac{\sigma-\beta}{K_g} + \frac{\beta}{K_f}\right)\Big/\sigma\right]^{-1},$$

2.9 $$M = C/\sigma$$

where

2.10 $$\sigma = 1 - K^*/K_g.$$

Equation (2.7) is Gassmann's equation [20]. The derivation of (2.7) requires the additional assumption that the saturated porous medium is homogeneous on the microscopic scale as well as on the macroscopic scale. Brown and Korringa [21] have shown that Gassmann's equation can be generalized by removing the assumption of microhomogeneity. Korringa [16] has shown recently that (2.7)-(2.9) may also be generalized by removing this assumption. The resulting formulas have the same form as (2.7)-(2.9) but C and the parameter σ are no longer given by (2.8) and (2.10). For our present purposes, we will retain the assumption of microhomogeneity and use (2.7)-(2.10).

To complete the model for the coefficients, either experimental or theoretical estimates of the frame moduli are required. It has been pointed out recently [5,7] that K^* and μ^* may be estimated theoretically using effective medium theory for elastic composites. Since the moduli K^* and μ^* are moduli of the (dry) porous frame, it follows that these moduli can be estimated by treating the frame as a two phase composite: one phase is a vacuum $(K_1, \mu_1, f_1) = (0, 0, \beta)$ while the other phase consists of the solid grains $(K_2, \mu_2, f_2) = (K_g, \mu_g, 1-\beta)$. The effective medium theory [7-10] then predicts that the moduli can be estimated using the coupled equations (summation convention assumed for i, j)

2.11
$$\sum_{n=1}^{2} f_n(K_n - K^*) \frac{1}{3} T_{iijj}^{(n)} = 0$$

and

2.12
$$\sum_{n=1}^{2} f_n(\mu_n - \mu^*) \frac{1}{5} (T_{ijij}^{(n)} - \frac{1}{3} T_{iijj}^{(n)}) = 0$$

where the f_n's are the volume fractions. The values of the scalars $T_{iijj}^{(2)}$ and $T_{ijij}^{(2)}$ depend on K_g, μ_g, K^*, μ^* and also on the assumed shapes of the inclusions. The scalars have been computed for ellipsoidal inclusions by Wu[22,10]. We have obtained good results by assuming the pores are approximated by needle-shaped inclusions [5]. Other authors prefer to use different ellipsoidal shapes to approximate the pore space [7].

If the strain energy functional is written in the form

2.13
$$E = \frac{1}{2}(\tau_{ij} e_{ij} + p\zeta)$$

where p is the pressure in the fluid and τ_{ij} is the average stress tensor for the saturated porous medium, it follows that the Fourier transformed equations of motion for elastic waves take the form

2.14
$$\tau_{ij,j} = (H-\mu^*)e_{,i} + \mu^* u_{i,jj} - C\zeta_{,i}$$
$$= -\omega^2(\rho u_i + \rho_f w_i)$$

2.15
$$-p_{,i} = Ce_{,i} - M\zeta_{,i}$$
$$= -\omega^2(\rho_f u_i + q w_i).$$

A subscript following a comma indicates a derivative and we have assumed a time dependence of the form $\exp(-i\omega t)$. The new coefficients in (2.14) and (2.15) are the average density

2.16
$$\rho = (1-\beta)\rho_g + \beta\rho_f$$

and

2.17
$$q(\omega) = \alpha\rho_f/\beta + iF(\xi)\nu\rho_f/\kappa\omega$$

where

2.18
$$\xi = (\omega a^2/\nu)^{1/2}.$$

The kinematic viscosity of the fluid is ν; the permeability of the porous frame is κ; the parameter a is some characteristic length. The factors α and $F(\xi)$ are dimensionless numbers of magnitude greater than or equal to unity which depend on the geometry of the porous material. To gain insight into the general form of α and F, it is instructive to consider some limiting cases where exact results are known.

In the high porosity limit, Chase [23] has derived equations equivalent to (2.14) and (2.15) with the factor q given by

2.19 $\qquad q(\omega) = \rho_f [1 + i\Delta(\omega)]/\beta$

where the drag-interaction coefficient Δ for spherical particles of radius a oscillating in the fluid is given by [24]

2.20 $\qquad \Delta(\omega) = \frac{9}{4} (\beta^{-1} - 1)[1 + z - iz(1 + 2z/9)]z^{-2}$

where $z = 2^{-1/2}\xi$. Equating (2.17) and (2.19), we find

2.21 $\qquad \alpha = 1 + \frac{1}{2}(\beta^{-1} - 1)$

and, if we take the permeability of a suspension of spheres to be $\kappa = 2a^2\beta^2/9(1-\beta)$ [25,26], then

2.22 $\qquad F(\xi) = 1 + 2^{-1/2}\xi(1-i).$

The result (2.21) for the structure factor agrees with the general form $\alpha = 1 + r(\beta^{-1} - 1)$, where $r = \frac{1}{2}$ for spheres and $0 \leq r \leq 1$ for arbitrary ellipsoids, derived in [5]. Equation (2.22) generalizes that result to include frequency dependent effects.

Equations (2.21) and (2.22) are reasonable estimates of α and $F(\xi)$ in the high porosity limit. For lower porosity, we may expect significant deviations from (2.21) and (2.22) to occur. To obtain an estimate of the structure factor for all porosities, consider a recent result of Brown [11] showing that

2.23 $\qquad \alpha = \beta F$

where F is the low-frequency electrical-resistivity formation factor. Furthermore, Sen, Scala, and Cohen [13] have recently shown both theoretically and experimentally that, for spherical glass beads,

2.24 $\qquad F = \beta^{-3/2}$

to a very good approximation over a wide range of porosities ($.01 < \beta < 1$). Combining (2.23) and (2.24) gives the estimate

2.25 $\quad \alpha = \beta^{-1/2}$.

Defining $\delta = 1 - \beta$ and comparing the Taylor series expansion of (2.21) and (2.25) at high porosity ($\delta \ll 1$), we find in both cases that

2.26 $\quad \alpha \tilde{=} 1 + \frac{1}{2} \delta$

for small δ. We conclude that either (2.21) or (2.25) may be used at high porosity but that (2.25) is preferred for lower porosity materials.

To obtain estimates of $F(\xi)$ valid at lower porosities, Biot [2] did two model calculations to determine the effective frequency dependence of the fluid viscosity (1) between parallel plates and (2) in cylindrical tubes. He also discussed the deviations from these forms for other pore shapes. In both model calculations, Biot found functions whose limiting forms are similar to (2.22). In particular, for cylindrical tubes of radius a and with our choice of sign for the frequency dependence Biot's result is

2.27 $\quad F(\xi) = \frac{1}{4} \frac{\xi T(\xi)}{1 + 2T(\xi)/i\xi}$

where

2.28 $\quad T(\xi) = \frac{\text{ber}'(\xi) - i\, \text{bei}'(\xi)}{\text{ber}(\xi) - i\, \text{bei}(\xi)}$.

The functions $\text{ber}(\xi)$ and $\text{bei}(\xi)$ are the real and imaginary parts of the Kelvin function. The limiting forms of (2.27) are

2.29 $\quad F(\xi) \to 1 \qquad$ for $\xi \ll 1$ and

2.30 $\quad F(\xi) \to 2^{-1/2} \xi(1-i)/4 \quad$ for $\xi \gg 1$.

The corresponding permeability for a porous medium with straight capillaries of radius a is $\kappa = \beta a^2/8$. Except for a factor of 4 at high frequencies, the limiting forms (2.29) and (2.30) agree with the limiting forms of (2.22) but the interpretation of the characteristic length a is different in the two cases: a is a typical grain radius in (2.22) while it is a typical pore radius in (2.27).

Biot [2] postulates that there exists a universal function $F(\xi)$. Since the limiting forms of (2.22) and (2.27) (and also the result of the parallel plate calculation) have the same fundamental form, Biot's postulate may be correct. However, to establish the form of this universal function requires a detailed calculation based on the microstructure. Such a calculation may be possible, e.g. using the formalism of Burridge and Keller [27], but so far such a calculation has not been done. To complete our model, we will instead assume that (2.27) gives a reasonable approximation to the universal function for some choice of a and treat the characteristic length a as a model parameter. Ideally, we should have reliable theoretical estimates for both the permeability κ and the characteristic length a. Although some progress has been made in obtaining consistent estimates of κ and a [28], at present no model known to the author is capable of making the transition from the high porosity limit to the low porosity limit in a completely satisfactory manner. The particular choices we make for the permeability and characteristic length will be discussed in the next section along with the comparison to recent experimental results.

3. Wave speeds and attenuation

One result which is easily derived from the coupled equations (2.14)-(2.15) is the existence of three distinct modes of elastic wave propagation: two compressional waves and one shear wave. The phase speed of the shear mode is given by

3.1 $$v_s^2 = \mu^*/(\rho - \rho_f^2/q).$$

The coupled equations for the two compressional wave speeds v_+ and v_- are given by the determinant

3.2 $$\begin{vmatrix} \rho - Hv^{-2} & \rho_f - Cv^{-2} \\ \rho_f - Cv^{-2} & q - Mv^{-2} \end{vmatrix} = 0.$$

It is known [1,29] that the two compressional modes correspond to an in-phase motion of the fluid and solid at speed v_+ and a second, out-of-phase, motion at speed v_-. These modes generally correspond to dispersive waves although the dispersion is weak at ultrasonic frequencies. In particular, the second compressional mode is strongly attenuated at low frequencies, having the character of a diffusion process.

In recent experiments, Plona [4] observed two distinct compressional waves in a water-saturated, porous structure composed of sintered glass beads. It has been shown [5,6] that Plona's data for the wave speeds are consistent with the predictions of Biot's theory.

The input parameters to our model [5] are K_g = 0.407 Mb, μ_g = 0.297 Mb, ρ_g = 2.48 g/cc, K_f = 0.022 Mb, ρ_f = 1.00 g/cc, ν = 1.00 centistoke, and $\omega = 2\pi \times 500$ kHz. The frame moduli K^* and μ^* are calculated from (2.11)-(2.12) assuming needle-shaped inclusions. For the present comparison, we use (2.25) instead of (2.21) as our estimate of the structure factor. We let the permeability obey the Kozeny-Carman relation [28].

3.3 $$\kappa(1-\beta)^2/\beta^3 = \kappa_o(1-\beta_o)^2/\beta_o^3 = \text{const}$$

which has been shown empirically to provide reasonable estimates of the dependence of permeability on porosity. We choose $\kappa_o = 9.1 \times 10^{-8}$ cm^2 and $\beta_o = 0.283$ to agree with one of Plona's measured values [4,30] of the permeability and use (3.3) to predict the value of κ for all other porosities considered. No entirely satisfactory model for the characteristic length a has been found. However, dimensional analysis suggests that a^2 must be comparable to κ, so we take

3.4 $$a^2/\kappa = a_o^2/\kappa_o = \text{const.}$$

Although this ratio is certainly a function of porosity [28], we expect the porosity dependence to be rather weak for moderate values of the porosity. At $\beta_o = 0.283$, we choose $a_o = 0.02$ mm corresponding to an average pore radius [15] $\frac{1}{5}$ to $\frac{1}{7}$ of the grain radius (the glass beads were 0.21-0.29 mm in diameter).

The speeds predicted by this model are compared to Plona's observed values in Figure 1. The theoretical results for the fast compressional wave and the shear wave agree with Plona's measurements within the experimental error (3% relative error in measured speeds and an absolute error of ±0.005 in measured porosity). Agreement is not quite so good for the slow compressional wave. The values at $\beta = 0.283$ and 0.258 are within the error bars but the value at $\beta = 0.185$ is higher than the measured value by about 10%. Nevertheless, we consider the agreement to be very good.

Biot's theory can also be used to estimate the ratio of the direct slow wave peak amplitude to the direct fast wave peak amplitude at normal incidence. To do so, we must consider the boundary conditions appropriate for an interface between a fluid and a fluid-saturated porous medium. Assuming free flow across the interface and that the exterior and interior fluids are the same, Deresiewicz and Skalak [31] have shown that the appropriate boundary conditions are

3.5 $$\tau_{nn} = -P,$$

3.6 $$\tau_{ns} = 0,$$

3.7 $$p = P, \quad \text{and}$$

3.8 $$u_{n,t} + w_{n,t} = U_{n,t},$$

where τ_{nn} (no summation convention) and τ_{ns} are the normal and shear components of the average stress and P is the pressure of the exterior fluid. The final equation (3.8) follows from the continuity of the velocity normal to the

Figure 1. Comparison of theoretical predictions and experimental results for speeds of the fast compressional wave (v_+), the shear wave (v_s), and the slow compressional wave (v_-). Input parameters to the model are: $K_g = 0.407$ Mb, $\mu_g = 0.297$ Mb, $\rho_g = 2.48$ g/cc, $K_f = 0.022$ Mb, $\rho_f = 1.00$ g/cc, $\nu = 0.01$ cm^2/sec, $\omega = 2\pi \times 500$ kHz, $\kappa_o = 9.1 \times 10^{-8}$ cm^2, $a_o = 0.02$ mm, and $\beta_o = 0.283$.

interface averaged over the bulk area, $U_{n,t}$ being the normal component of the velocity of the exterior fluid.

Lack of space precludes us from presenting the details of the calculation for the amplitude ratios here. However, a brief discussion of the results of the analysis is appropriate. Assuming a plane pressure-wave of unit amplitude and frequency ω incident on a slab of saturated porous material of thickness d, we can compute the amplitudes of the fast and slow (mode converted) compressional waves transmitted into the medium. For the three largest porosities in Plona's published data, we find the amplitude of the slow wave is only about 1-3% of that of the fast wave. For $\beta = 0.075$, this ratio is less than 0.3%. When these waves reach the second interface, we find that a large fraction of the energy in the fast wave is reflected while the energy in the slow wave is much more strongly converted and transmitted into the exterior fluid. Thus, if attenuation of the ultrasonic pulses could be neglected as they traverse the porous sample, two transmitted signals of comparable amplitude would be received at a detector. If the wavenumber of a particular mode is $k = \omega/v$, then the attenuation of amplitude for that mode in a sample of thickness d is a factor of the form $\exp(-|\text{Im} k|d)$. If, in addition to the effects already described, we include attenuation of both compressional waves in the calculation, we find the results displayed in Table 1. (Note that the experimental error in the measured amplitude ratios is typically 10%). The attenuation of the fast wave is nearly negligible in all four samples so that, except for a factor close to unity, Biot's theory predicts that the observed amplitude ratio is due to the attenuation of the slow wave.

It should be stressed that small differences in the assumed value of permeability lead to large differences in the computed attenuation of the slow wave. It is therefore important that accurate measurements of the permeability for these particular samples be made.

We see that the theory predicts amplitude ratios in good agreement with experiment for the porosities $\beta = 0.283$ and 0.185. For $\beta = 0.075$, the theory predicts an amplitude ratio of 0.02, which is so low that the experimental technique used would not be able to detect a slow wave of this amplitude; hence, the theory explains why no slow wave was observed in this case. The case with $\beta = 0.258$ is therefore the only one where theory and experiment disagree significantly. This discrepancy could be due to an anomalously small permeability for this sample; further experiments are required to resolve this issue.

The theory is in sufficiently good agreement with the data that we may conclude (1) that Biot's theory is a satisfactory model of wave speeds and attenuation in fluid-saturated porous media and (2) that more experiments are

Sample	1	2	3	4
Porosity	0.283	0.258	0.185	0.075
Thickness (cm)	1.880	1.793	1.646	1.461
Observed amplitude ratio ($\theta=0$)	0.38	0.26	0.21	---
Predicted amplitude ratio ($\theta=0$)	0.35	0.32	0.19	0.02

Table 1. Observed and predicted slow wave to fast wave amplitude ratios for first arrivals at normal incidence through water-saturated sintered glass samples of the stated porosity (β) and thickness (d). All experimental data courtesy of T. J. Plona.

needed to answer the questions raised by the few remaining anomalies. Further theoretical work is needed to find the most general form of the universal function $F(\xi)$ and to establish more satisfactory estimates of the characteristic length a. More experimental work is needed to check the dependence of wave attenuation on sample permeability and to determine whether Biot's theory correctly predicts the reflection and transmission characteristics of waves incident at non-normal angles.

References

[1] M. A. Biot, Theory of propagation of elastic waves in a fluid-saturated porous solid. I. Low-frequency range, J. Acoust. Soc. Am. $\underline{28}$, 168 - 178 (1956).

[2] M. A. Biot, Theory of propagation of elastic waves in a fluid-saturated porous solid. II. Higher frequency range, J. Acoust. Soc. Am. $\underline{28}$, 179 - 191 (1956).

[3] M. A. Biot, Mechanics of deformation and acoustic propagation in porous media, J. Appl. Phys. $\underline{33}$, 1482 - 1498 (1962).

[4] T. J. Plona, Observation of a second bulk compressional wave in a porous medium at ultrasonic frequencies, Appl. Phys. Lett. $\underline{36}$, 259 - 261 (1980).

[5] J. G. Berryman, Confirmation of Biot's theory, Appl. Phys. Lett. $\underline{37}$, 382 - 384 (1980).

[6] N. C. Dutta, Theoretical analysis of observed second bulk compressional wave in a fluid-saturated porous solid at ultrasonic frequencies, Appl. Phys. Lett. $\underline{37}$, 898 - 900 (1980).

[7] J. Korringa, R. J. S. Brown, D. D. Thompson, and R. J. Runge, Self-consistent imbedding and the ellipsoidal model for porous rocks, J. Geophys. Res. $\underline{84}$, 5591 - 5598 (1979).

[8] J. G. Berryman, Theory of elastic properties of composite materials, Appl. Phys. Lett. $\underline{35}$, 856 - 858 (1979).

[9] J. G. Berryman, Long-wavelength propagation in composite elastic media I. Spherical inclusions, J. Acoust. Soc. Am. $\underline{68}$, 1809 - 1819 (1980).

[10] J. G. Berryman, Long-wavelength propagation in composite elastic media II. Ellipsoidal inclusions, J. Acoust. Soc. Am. $\underline{68}$, 1820 - 1831 (1980).

[11] R. J. S. Brown, Connection between formation factor for electrical resistivity and fluid-solid coupling factor in Biot's equations for acoustic waves in fluid-filled porous media, Geophysics $\underline{45}$, 1269 - 1275 (1980).

[12] D. L. Johnson, Equivalence between fourth sound in liquid He II at low temperatures and the Biot slow wave in consolidated porous media, Appl. Phys. Lett. $\underline{37}$, 1065 - 1067 (1980).

[13] P. N. Sen, C. Scala, and M. H. Cohen, A self-similar model for sedimentary rocks with application to the dielectric constant of fused glass beads, Geophysics $\underline{46}$, 781 - 795 (1981).

[14] R. D. Stoll and G. M. Bryan, Wave attenuation in saturated sediments, J. Acoust. Soc. Am. $\underline{47}$, 1440 - 1447 (1970).

[15] R. D. Stoll, Acoustic waves in saturated sediments, in *Physics of Sound in Marine Sediments,* edited by L. Hampton (Plenum, New York, 1974), pp. 19 - 39.

[16] J. Korringa, On the Biot-Gassmann equations, to appear.
[17] A. E. H. Love, *A Treatise on the Mathematical Theory of Elasticity* (Dover, New York, 1944), pp. 43, 62, 102.
[18] M. A. Biot and D. G. Willis, The elastic coefficients of the theory of consolidation, J. Appl. Mech. $\underline{24}$, 594 - 601 (1957).
[19] J. Geertsma, The effect of fluid pressure decline on volumetric changes of porous rocks, Trans. AIME $\underline{210}$, 331 - 340 (1957).
[20] F. Gassmann, Über die elastizität poröser medien, Veirteljahrsschrift der Naturforschenden Gesellschaft in Zürich $\underline{96}$, 1 - 23 (1951).
[21] R. J. S. Brown and J. Korringa, On the dependence of the elastic properties of a porous rock on the compressibility of the pore fluid, Geophysics $\underline{40}$, 608 - 616 (1975).
[22] T. T. Wu, The effect of inclusion shape on the elastic moduli of a two-phase material, Int. J. Solids Structurers $\underline{2}$, 1 - 8 (1966).
[23] D. M. Chase, Wave propagation in liquid-saturated open-cell foam, J. Acoust. Soc. Am. $\underline{65}$, 1 - 8 (1979).
[24] L. D. Landau and E. M. Lifshitz, *Fluid Mechanics* (Pergamon Press, London, 1959), p. 96.
[25] T. S. Lundgren, Slow flow through stationary random beds and suspensious of spheres, J. Fluid Mech. $\underline{51}$, 273 - 299 (1972).
[26] J. M. Hovem, Viscous attenuation of sound in suspensions and high-porosity marine sediments, J. Acoust. Soc. Am. $\underline{67}$, 1559 - 1563 (1980).
[27] R. Burridge and J. B. Keller, Poroelasticity equations derived from microstructure, to appear.
[28] J. M. Hovem and G. D. Ingram, Viscous attenuation of sound in saturated sand, J. Acoust. Soc. Am. $\underline{66}$, 1807 - 1812 (1979).
[29] J. Geertsma and D. C. Smit, Some aspects of elastic wave propagation in fluid-saturated porous solids, Geophysics $\underline{26}$, 169 - 181 (1961).
[30] T. J. Plona and D. L. Johnson, Experimental study of the two bulk compressional modes in water-saturated porous structures, in *Proceedings of the November, 1980, IEEE Ultrasonics Symposium*, to appear.
[31] H. Deresiewicz and R. Skalak, On uniqueness in dynamic poroelasticity, Bull. Seismol. Soc. Am. $\underline{53}$, 783 - 788 (1963).

BIOT'S POROELASTICITY EQUATIONS BY HOMOGENIZATION[*]

Robert Burridge
Courant Institute of Mathematical Sciences, New York University
New York, NY 10012

and

Joseph B. Keller
Departments of Mathematics and Mechanical Engineering
Stanford University
Stanford, CA 94305

Abstract

Equations are derived which govern the linear macroscopic mechanical behavior of a porous elastic solid saturated with a compressible viscous fluid. The derivation is based on the equations of linear elasticity in the solid, the linearized Navier-Stokes equations in the fluid, and appropriate conditions at the solid-fluid boundary. The scale of the pores is assumed to be small compared to the macroscopic scale, so that the two-space method of homogenization can be used to deduce the macroscopic equations. When the dimensionless viscosity of the fluid is small, the resulting equations are those of Biot, who obtained them by hypothesizing the form of the macroscopic constitutive relations. The present derivation verifies those relations, and shows how the coefficients in them can be calculated, in principle, from the microstructure. When the dimensionless viscosity is of order one, a different equation is obtained, which is that of a viscoelastic solid.

1. Introduction

Poroelasticity, the mechanics of porous elastic solids with fluid-filled pores, has received attention recently for several reasons: pore fluids in the ground are believed to play a role in the triggering of earthquakes; liquid waste disposed of underground seeps into pores; the enhanced recovery of oil, gas, and geothermally heated water depends upon flow in porous strata; underwater acoustics involves propagation in the water-saturated porous bottom of the ocean; etc. The poroelastic equations derived by Biot [1-6] have long been regarded as standard and have formed the basis for solving particular problems in

---

[*] Research supported in part by the Office of Naval Research, the Air Force Office of Scientific Research, the Army Research Office, and the National Science Foundation, Divisions of Mathematics and Earth Sciences. Presented at O.N.R. Conference on Ocean Bottom Effects; Miami, April 1979.

poroelasticity, e.g. Rice and Cleary [7], Burridge and Vargas [8]. Recently, however, the validity of these equations has been questioned, in particular by Cleary [9,10].

In view of their importance, and because of doubts about them, we have derived anew the governing equations of linear poroelasticity by starting with the detailed microstructure of the pores, the linearized equations of elasticity and the linearized equations of fluid dynamics. The equations we obtain coincide with Biot's provided that the dimensionless viscosity of the fluid is small. Our analysis also confirms the form of the constitutive relations which he assumed. In addition it shows how the coefficients in those relations can be calculated in principle from the microstructure of the medium. However when the dimensionless viscosity of the pore fluid is large, we obtain a different equation and different constitutive relations. They characterize the porous medium as a viscoelastic solid, in agreement with the suggestion of Cleary [9,10]. In this case also we show how to calculate the coefficients in the equation and in the constitutive relations in terms of the microstructure.

Our derivation is based upon the two-space method of homogenization, which was developed and studied by Sanchez-Palencia [11], Keller [12], Bensoussan, Lions and Papanicolaou [13], Larsen [14], and others. It applies to media with microstructure on a scale much smaller than the macroscopic scale of interest. The method provides a systemative procedure for deriving macroscopic dynamical equations from the basic dynamical equations which govern the behavior of the medium on the microscale. Thus it involves an appropriate form of averaging to eliminate the rapid variations associated with the small scale. It applies both to linear problems, such as we consider here, and to nonlinear problems. Sanchez-Palencia [11] and his co-workers have also derived the poroelastic equations by the two-space method, but their results were not identified with Biot's equations. Keller [15] used it to treat finite velocity flow of a compressible fluid through a rigid porous medium. Lehner [16] used a different method to derive Darcy's law for the flow of an incompressible fluid in a rigid porous medium.

In this summary we shall merely list our starting equations and results. For full details of the derivation see Burridge and Keller [17].

2. Formulation of the equations

We consider an elastic solid permeated by pores containing a compressible viscous fluid. We assume that the pore configuration has a scale length h which is small compared to a typical macroscopic scale length H. The material is fine grained, and appears locally homogeneous on the macroscopic scale. The small ratio h/H plays a key role in our analysis; we denote it by ε:

$$\varepsilon = h/H . \qquad (2.1)$$

To make precise the fact that the medium varies rapidly on the small scale h and may also vary slowly on the larger scale H, we assume that every property of the medium is of the form f(x,y) where

$$y = \varepsilon^{-1} x . \qquad (2.2)$$

Here $x = (x_1, x_2, x_3)$ is the position vector of a point in cartesian coordinates and $y = (y_1, y_2, y_3)$ is the vector of stretched coordinates. Thus y varies by an amount of order H as x varies by the small amount h. The first argument in f(x,y) accounts for the slow or gradual variation of any material property, and the second argument accounts for the rapid variation. For example f(x,y) may be a periodic function of y, an almost periodic function of y, a stationary random function of y, etc. In particular the characteristic or indicator function $\chi_s(x,y)$ of the solid region D_s is of this form. It is defined to have the value $\chi_s = 1$ at a point in the solid region D_s and the value $\chi_s = 0$ at a point outside D_s. Since we assume that the pores are filled with fluid, a point of the medium outside the solid region D_s is in the fluid region D_f, so $\chi_s = 0$ at a point in D_f.

We consider motions of the fluid and solid which are small enough to be governed by the linearized Navier-Stokes equations and the linearized equations of elasticity respectively, together with linearized interface conditions. For time harmonic motions with angular frequency ω, these equations are:

$$i\omega \rho_f v = \nabla \cdot \sigma , \qquad \text{in } D_f , \qquad (2.3a)$$

$$\sigma = -pI + \tilde{\mu} D \nabla v \qquad \text{in } D_f , \qquad (2.3b)$$

$$i\omega p = -\kappa \nabla \cdot v \qquad \text{in } D_f , \qquad (2.3c)$$

$$v = i\omega u \qquad \text{on } \partial D_f = \partial D_s , \qquad (2.3d)$$

$$n \cdot \sigma = n \cdot \tau , \qquad \text{on } \partial D_f = \partial D_s , \qquad (2.3e)$$

$$-\omega^2 \rho_s u = \nabla \cdot \tau, \quad \text{in } D_s, \qquad (2.3f)$$

$$\tau = C\nabla u, \quad \text{in } D_s. \qquad (2.3g)$$

Here the fluid occupies the domain D_f and has velocity v, pressure p, stress tensor σ, density ρ_f, viscosity $\tilde{\mu}$, and bulk modulus κ. The solid occupies the domain D_s and has displacement u, density ρ_s and elastic constants given by the fourth rank tensor $C(x,y)$ which operates on ∇u to yield the stress tensor τ, and n is the unit normal to $\partial D_f = \partial D_s$ pointing into the solid. In (2.3b) D is the operator defined by

$$De = \frac{1}{2}(e + e^T - \frac{2}{3} I \operatorname{tr} e). \qquad (2.4)$$

Thus De is the symmetrized deviatoric part of the second rank tensor e.

We consider the unknown field quantities $v(x,y), p(x,y), u(x,y), \sigma(x,y)$ and $\tau(x,y)$ to be functions of the two sets of space variables, just as we did for the material properties. However we consider these two sets of variables to be independent, so we are extending the definitions of the field quantities from three to six-dimensional space. Then the actual physical values of the field quantities are obtained by restricting these functions to the physical diagonal defined in (2.2). Also we extend the equations (2.3) by replacing ∇ by $\nabla_x + \varepsilon^{-1}\nabla_y$. This replacement is made because on the physical diagonal, where y is related to x by (2.2), we have

$$\nabla f(x,y) = \nabla f(x,\varepsilon^{-1}x) = \nabla_x f + \varepsilon^{-1} \nabla_y f. \qquad (2.5)$$

We assume that the dimensionless viscosity $\tilde{\mu}/\omega \rho_f h^2$, appropriate to the small scale, is of order unity. Therefore the dimensionless viscosity appropriate to the large scale, $\tilde{\mu}/\omega \rho_f H^2$ is small of order ε^2. To make this evident we replace $\tilde{\mu}$ by $\varepsilon^2 \mu$ in (2.3). In this case we ultimately obtain equations similar to those introduced by Biot [1-6]. We also treat the case in which $\tilde{\mu}/\omega \rho_f H^2$ us of order unity (see [17]).

We write each field quantity f in the form

$$f(x,y,\varepsilon) = f_0(x,y) + \varepsilon f_1(x,y) + \frac{\varepsilon^2}{2} f_2(x,y) + o(\varepsilon^2). \qquad (2.6)$$

where the $f_i(x,y)$ are bounded as functions of y. This procedure groups the terms of equations (2.3) into different orders (powers) of ε. We seek equations governing the leading coefficients $v_0, \sigma_0, p_0, u_0, \tau_0$. To accomplish this it is necessary to consider also terms of first order in ε but these are eliminated by an averaging process over the fast variable y. See [17] for details.

3. Final equations and comparison with Biot's equation

Denoting a spatial average in the y variable by an overbar our final equations may be written

$$-\omega^2 (\bar{\rho} u_0 + \rho_f \bar{w}) = \nabla_x \cdot (\bar{\tau}_0 - V_f p_0 I) , \tag{3.1a}$$

$$-\omega^2 \rho_f u_0 + \bar{W}^{-1} \bar{w} = -\nabla_x p_0 . \tag{3.1b}$$

$$\bar{\tau}_0 - V_f p_0 I = [\bar{C} + \overline{CLC} + (V_f I - \overline{CQ}) M \, \mathrm{tr} (V_f - \overline{LC})] \nabla_x u_0$$
$$+ (V_f I - \overline{CQ}) M \nabla_x \cdot \bar{w} . \tag{3.1c}$$

$$p_0 = -M \, \mathrm{tr} (V_f - \overline{LC}) \nabla_x u_0 - M \nabla_x \cdot \bar{w} . \tag{3.1d}$$

where the operators Q, L, W are defined in [17]. V_f, V_s are the fluid and solid volume fractions.

However, Biot [6] gives the poroelastic equations of motion in the form

$$\frac{\partial \tau_{ij}}{\partial x_j} = \rho \ddot{u}_i + \rho_f \ddot{w}_i , \tag{3.2a}$$

$$-\frac{\partial p_f}{\partial x_i} - \rho_f \ddot{u}_i = \bar{Y}_{ij}(p) \dot{w}_j , \tag{3.2b}$$

$$\tau_{ij} = A_{ij}^{\mu\nu} e_{\mu\nu} + M_{ij} \zeta , \tag{3.2c}$$

$$p_f = M_{ij} e_{ij} + M \zeta . \tag{3.2d}$$

It will be seen that our equations (3.1) agree exactly with (3.2) if the following identifications are made:

Present Work	Biot [6]	
$\bar{\rho}, \rho_f, M$	ρ, ρ_f, M	(3.3a)
$\bar{C} + \overline{CLC} + (V_f I - \overline{CQ}) M \, \mathrm{tr} (V_f - \overline{LC})$	$(A_{ij}^{\mu\nu})$	(3.3b)
$i\omega$	$i\omega = p = d/dt$	(3.3c)

Present Work	Biot [6]	
$\overline{W}^{-1}$	$p\overline{Y}_{ij}(p)$	(3.3d)
u_0, $\overline{w}$, p_0	u, w, p_f	(3.3e)
$M(V_f I - \overline{CQ}) = M\,\text{tr}(V_f I - \overline{LC})$	$(-M_{ij})$	(3.3f)
$\frac{1}{2}[\nabla_x u_0 + (\nabla_x u_0)^T]$, $\nabla \cdot \overline{w}$	e_{ij}, $-\zeta$	(3.3g)

Not only do our equations agree with Biot's in form, but the identification (3.3) gives a detailed prescription (see [17]) for calculating the coefficients from the micro-structure. Some of these are merely averages, such as the mass density of the bulk material $\overline{\rho} = \overline{\rho}_s + \overline{\rho}_f$, while others are far from obvious, such as (3.3b). Notice that in defining averages over quantities defined in $D_{s,f}$ an integration over $D_{s,f}$ was performed but the result was divided by the full bulk volume. However our quantities L, Q and W can be obtained only after the solution of special problems in elasticity or in fluid mechanics. They could be solved numerically for media with prescribed periodic microstructure, for instance.

When $\tilde{\mu}$ of (2.3b) is taken to be of order 1, rather than of order ε^2, we do not obtain Biot's equations but find that the bulk material behaves like a viscoelastic solid. For further details the reader is referred to [17].

References

1. M.A. Biot, General theory of three-dimensional consolidation, J. App. Phys. **12**, 155-164 (1941).

2. _____, Theory of elasticity and consolidation for a porous anisotropic solid, J. App. Phys. **26**, 182-185 (1955).

3. _____, General solutions of the equations of elasticity and consolidation for a porous material, J. App. Mech. **23**, 91-95 (1956).

4. _____, The theory of propagation of elastic waves in a fluid-saturated porous solid, I. Low frequency range, II. Higher frequency range, J. Acoust. Soc. Am. **28**, 168-178, 179-191 (1956).

5. _____, Mechanics of deformation and acoustic propagation in porous media, J. App. Phys. **33**, 1482-1498 (1962).

6. _____, Generalized theory of acoustic propagation in porous dissipative media, J. Acoust. Soc. Am. **34**, 1256-1264 (1962).

7. J.R. Rice and M.P. Cleary, Some basic stress-diffusion solutions for fluid saturated elastic porous media with compressible constituents, Rev. Geophys. Space Phys. **14**, 227-241 (1976).

References, continued

8. R. Burridge and C.A. Vargas, The fundamental solution in dynamic poroelasticity, Geophys. J. Roy. Astro. Soc. **58**, 61-90 (1978).

9. M.P. Cleary, Fundamental solutions for fluid-saturated porous media and applications to localized rupture phenomena, Ph.D. Thesis, Brown University, Division of Engineering, 1975.

10. _____, Elastic and dynamic response regimes of fluid-impregnated solids with diverse microstructures, Int. J. Solids Structures **14**, 795-819 (1978).

11. E. Sanchez-Palencia, Non-homogeneous Media and Vibration Theory, Lecture Notes in Physics **127**, Springer-Verlag, 1980.

12. J.B. Keller, Effective behavior of heterogeneous media, in <u>Statistical Mechanics and Statistical Methods in Theory and Application</u>, U. Landman, ed., Plenum, New York, 1977, pp. 631-644.

13. A. Bensoussan, J.-L. Lions, and G.C. Papanicolaou, Asymptotic analysis for periodic structures, in <u>Studies in Mathematics and Its Applications</u>, Vol. 5, North-Holland, 1978.

14. E.W. Larsen, Neutron transport and diffusion in inhomogeneous media, II, Nucl. Sci. Eng. **60**, 357-368 (1976).

15. J.B. Keller, Darcy's law for flow in porous media and the two-space method, in <u>Nonlinear Partial Differential Equations in Engineering and Applied Science</u>, R.L. Sternberg, A.J. Kalinowski and J.S. Papadakis, eds., Marcel Dekker, New York, 1980, 429-443.

16. F.K. Lehner, A derivation of the field equations for slow viscous flow through a porous medium, I & EC Fundamentals **18**, 41-45 (1979).

17. R. Burridge and J.B. Keller, Poroelasticity equations derived from microstructure, J. Acoust. Soc. Amer. **70**, 1140-1146 (1981).

APPROXIMATIONS OF BRINKMAN TYPE

Stephen Childress*
New York University
Courant Institute of Mathematical Sciences

1. Introduction

In 1947 Brinkman [1] introduced an approximate method for computing the permeability k of a bed of identical randomly placed rigid spheres. Brinkman's method is based upon an ingenious use of an "effective medium" and a single test sphere. Since k can be derived from the expected force experienced by a single sphere within the bed, and since such a test sphere will be surrounded by a "cloud" of other spheres, it is tempting to compute an approximate value of the expected force by surrounding the test sphere by a homogeneous porous medium having the as yet unspecified permeability k. By solving the Stokes equations for flow past a sphere in such a medium, a force $F(k)$ can be found, in which case the permeability is obtained by solving the algebraic equation

$$(1.1) \qquad k^{-1} = nF(k)$$

where n is the number density of spheres in the bed.

This approach, which can be modified to deal with more complicated random geometries, is similar in spirit to certain (quasi crystalline) approximations in multiple scattering theory [2,3,4], and may generally be compared with any example of "renormalization" in deriving bulk properties of random materials. The value of k obtained from (1.1) agrees quite well with experiments [5], even outside the range of validity one might expect on physical grounds, in spite of the fact that the only statistical parameter which appears is the number density n.

One unfortunate aspect of this approximation is that the error is difficult to estimate analytically. From the viewpoint of multiple scattering theory (1.1) is obtained from an infinite partial summation of a series which is formally divergent. From a practical standpoint, however, such estimates are useful since in many cases the statistics of the random material will be known only partially or imprecisely.

---

*Research supported by the National Science Foundation under Grant MCS-79-02766.

In the present paper we study Brinkman's approximation in a one-dimensional model problem which has many of the features of a permeability problem. In this model a number of questions can be studied rather easily: (i) How does Brinkman's formula compare with upper and lower bounds on permeability? (ii) What is the effect of statistical parameters other than n? (iii) How does the formula compare with Monte Carlo simulation with uniform statistics? (iv) Is there a hierarchy of Brinkman-type models of increasing accuracy? At the end of the paper we compare this study with some known results for the permeability of a random bed of spheres.

2. A One-dimensional Model Problem

We shall consider the mechanical system shown in Figure 1.

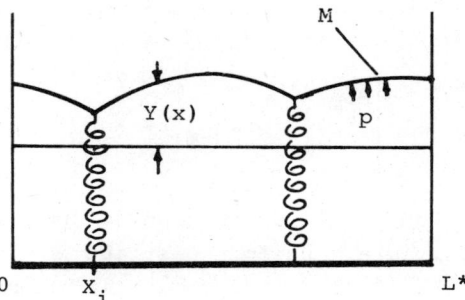

Figure 1. An elastic membrane is supported from below by a pressure p and attached at random points to linear springs.

An elastic membrane M, carrying constant tension T, is supported by a pressure difference p, and held by attachment at points X_i, i 1,2,$\cdots$,N in the interval (0,L*), to linear springs with spring constant K, the rest position of the point of attachment being on the x-axis. Outside the interval (0,L*) the membrane height Y(X) is extended as a continuous periodic function with period L*.

Assuming that the membrane slope remains small, the function Y(X) satisfies, for $-\infty < X < \infty$,

$$(2.1) \quad Y(X)'' = -p/T + k \sum_{i=1}^{N} \delta(X-X_i) Y(X_i), \qquad Y(X+L^*) = Y(X).$$

Since k and p/T are inverse lengths, it is convenient to adopt k^{-1} and $pT^{-1}k^{-2}$ as units of length in the horizontal and vertical directions, respectively, so that (2.1) can be written (with $Y = ypT^{-1}k^{-2}$,

$$X = xk^{-1}, \quad L = L*k^{-1}),$$

(2.2) $\quad y''(x) = -1 + \sum_{i=1}^{N} \delta(x-x_i) y(x_i), \quad y(x+L) = y(x).$

We wish now to consider (2.2) for arbitrary N, (x_i) and L.

Since the points are to be chosen randomly, we now let $(\tilde{x}_i)$ be an unordered sequence of points in $(0,L)$ and define $f_N(\tilde{x}_1, \ldots, \tilde{x}_N)$ to be the associated joint probability distribution and define the reduced distributions as

(2.3) $\quad f_m(\tilde{x}_1, \ldots, \tilde{x}_m) = \left(\frac{1}{L} \int_0^L\right)^{N-m} f_N d\tilde{x}_{N-m+1} \cdots d\tilde{x}_N,$

with f_1 normalized to be unity. We also assume the process to be stationary, so that all distribution functions are completely symmetric in all arguments. We then define

(2.4) $\quad <Q>_{1,2\cdots m} = \left(\frac{1}{L} \int_0^L\right)^{N-m} \frac{f_N}{f_m} Q d\tilde{x}_{N-m+1} \cdots d\tilde{x}_N$

as the conditional expectation of Q holding m points fixed. For the special case of a completely uniform distribution we have $f_m = 1$, all m. If N and L are allowed to approach infinity in such a way that $N/L = n$ has a nonzero limit, then it is known [6] that the sequence $(d_i = x_i - x_{i-1})$, where (x_i) is the ordering of $(\tilde{x}_i)$, constitutes a Poisson process, and this provides a convenient case for Monte Carlo simulation.

No doubt numerous probabilistic tools may be brought to bear to study various properties of the realizations of (2.2), but we shall confine attention to elementary estimates of the ratio

(2.5) $\quad \gamma = \frac{<y(\tilde{x}_1)>_1}{<y>}$

in the limit $N, L \to \infty$. We note that, upon integrating (2.2) over the interval we obtain

(2.6) $\quad 0 = -L + \sum_{i=1}^{N} y(\tilde{x}_i).$

If now we compute the expectation $<y>$ by multiplication with f_N and integration with respect to all x_i there results

(2.7) $\langle y \rangle'' = -1 + \frac{N-1}{L} f_1(x) \langle y(\tilde{x}_1) \rangle_1 = -1 + \frac{N-1}{L} \langle y(\tilde{x}_1) \rangle_1$.

But $\langle y \rangle$ is a constant and so (2.6) and (2.7) yield, if $N \gg 1$,

(2.8) $\langle y(\tilde{x}_1) \rangle_1 = \frac{1}{N} \sum_{i=1}^{N} y(\tilde{x}_i) = \frac{1}{n}$.

Thus γ^{-1} is just the expected mean deflection of the membrane (which intuitively should be small when the springs are distributed densely) times the number of springs per unit (K^{-1}) of length.

We remark that a related one-dimensional, reduced scattering problem has been studied in considerable detail by Bazer [7]. In that problem, the term -1 on the right of (2.2) is replaced by $-y$, and this change allows the probability distribution for y under Poisson statistics to be studied in a scalar Markov model [8]. This approach does not appear to be possible for (2.2).

<u>Brinkman's approximation</u>. If we think of $y(x)$ as an expectation and replace the distribution of springs by a continuous elastic bed, then $n\gamma y$, where γ as defined by (2.5) is the average spring elongation divided by y, determines the mean force exerted by the bed. Introducing a "test spring" at the origin we consider the problem

(2.9) $y'' = -1 + \delta(x) y(0) + n\gamma y$, $y \to \langle y \rangle$ as $x \to \infty$,

with solution

(2.10) $y = \langle y \rangle - y(0) (4n\gamma)^{-\frac{1}{2}} e^{-(n\gamma)^{\frac{1}{2}} |x|}$

From (2.10) we may compute the deflection $y(0)$ of the spring, and for this value to be self-consistent with the γ determining the continuous bed we must have

(2.11) $\frac{y(0)}{\langle y \rangle} = \left(1 + (4n\gamma)^{-\frac{1}{2}}\right) = \gamma$.

Thus

(2.12) $\gamma = \frac{1 + 8n - (1+16n)^{\frac{1}{2}}}{8n}$

is the Brinkman expression for $\gamma(n)$.

This procedure can be made somewhat more formal by considering the conditional expectation of (2.2) holding one point fixed. We have, upon multiplication by f_2/f_1 and integration with respect to $\tilde{x}_2, \tilde{x}_3, \cdots$, in the limit $N, L \to \infty$, $N/L \to n$,

(2.13) $\qquad <y>_1 = -1 + <y(\tilde{x}_1)>_1 \; \delta(x-\tilde{x}_1) + n<y(\tilde{x}_2)>_{12} \big|_{\tilde{x}_2=x}$,

and so (2.9) and (2.13) become equivalent with the assumption

(2.14) $\qquad <y(\tilde{x}_2)>_{12} \big|_{\tilde{x}_2=x} = \gamma <y(x)>_1$.

That is, the approximation (2.12) follows from the assumption that the one-point conditioned expectation of y responds to a second spring precisely as the unconditioned expectation, a step which clearly neglects any "interaction" between the two springs.

Since $1/n$ is the average spacing between uniformly distributed points, the solution (2.10), and the notion of a continuous elastic bed, would appear to be reasonable provided that $(n\gamma)^{-\frac{1}{2}} \gg n^{-1}$ or

(2.15) $\qquad (1+16n)^{\frac{1}{2}} \ll 1 + 4n$,

implying that n should be large.

<u>Upper and lower bounds on γ</u>. To compute an upper bound on γ for a given realization of points, we consider the class of functions $f(x)$ which are smooth on $(0, L)$ except for discontinuities in f' at the x_i, with periodic extension outside this interval, and with $\frac{1}{N} \sum_i f(x_i)$ equal to $1/n$. It is easily seen that the Euler-Lagrange equation for

(2.16) $\qquad F[f] = \frac{1}{2} \int_0^L f'^2 dx + \frac{1}{2} \sum_{i=1}^N f^2(x_i) - \int_0^L f dx$

is satisfied by the solution y of (2.2), and that

(2.17) $\qquad F[f] \geq F[y] = -L/2\gamma n$.

If we take

(2.18) $\qquad f = 1 - \lambda(x-x_{i-1})(x-x_i)$, $\qquad x_{i-1} < x < x_i$,

as trial function, (2.17) yields an optimal bound when $\lambda = \frac{1}{2}$:

$$(2.19) \quad \gamma \leq (1 + \beta_3/12n)^{-1}, \quad \beta_3 = n^3 \cdot \frac{1}{N} \sum_{i=1}^{N} d_i^3 .$$

For a Poisson distribution of the d_i we have, for large N,

$$(2.20) \quad \beta_3 = n^4 \int_0^\infty e^{-ns} s^3 ds = 3! .$$

To compute a lower bound, consider the class of functions $g(x)$ which satisfy $g' = -1$ except at the x_i, where discontinuities $[g]_i$ may occur, with again periodic extension outside $(0,L)$. It is then easily seen that

$$(2.21) \quad G[g] = -\frac{1}{2} \int_0^L g^2 dx - \frac{1}{2} \sum_{i=1}^{N} [g]_i^2$$

satisfies

$$(2.22) \quad G[g] \leq G(y) = -L/2\gamma n .$$

Choosing the trial function

$$(2.23) \quad g = x - d_i/2, \quad x_{i-1} < x < x_i ,$$

we then obtain

$$(2.24) \quad \gamma \geq (\beta_2 + \beta_3/12n)^{-1}, \quad \beta_2 = \frac{n^2}{4N} \sum_{i=1}^{N} (d_{i+1} + d_i)^2 .$$

For the uniform distribution and large N, $\beta_2 = 3/2$ and therefore the two bounds yield, for the uniform distribution,

$$(2.25) \quad (3/2 + 1/2n)^{-1} \leq \gamma \leq (1 + 1/2n)^{-1} .$$

An upper bound on $\langle\gamma\rangle$. Since the upper bound involves a relatively unconstrained variational problem, it is of interest to examine the possibility of a sharper result motivated by the form of Brinkman's test field (2.10). We therefore consider functions f^*, satisfying the conditions on the functions in (2.16) but now with $\langle f^*\rangle = 1/n$ replacing $\frac{1}{N} \sum_i f(x_i) = 1/n$. Let

$$(2.26) \quad F^*[f^*] = \frac{1}{2L} \int_0^L (f^{*\prime})^2 dx + \frac{1}{2L} \sum_{i=1}^N f^*(\tilde{x}_i))^2 .$$

Among such functions F^* is minimized by γy, where y solves (2.2), and we have

$$(2.27) \quad F^*[f^*] \geq F^*[\gamma y] = \gamma/2n .$$

As a trial function we now take

$$(2.28) \quad f^* = \frac{1}{n}\left[1 + A \sum_{i=1}^N \phi(x,\tilde{x}_i)\right] ,$$

where

$$(2.29) \quad \phi(x,\tilde{x}_i) = \cosh \mu(x-\tilde{x}_i) + \mathrm{sgn}(\tilde{x}_i-x)\tanh(\tfrac{1}{2}\mu L)\sinh \mu(x-\tilde{x}_i)$$

$$+ \frac{2}{\mu^2 L}\tanh(\tfrac{1}{2}\mu L) , \qquad 0 \leq x \leq L .$$

Note that this function satisfies $\phi_{xx} + \mu^2\phi = 0$ except at (x_i), and $\langle\phi\rangle = 0$. Introducing (2.29) into (2.27) and taking the expectation of γ we obtain:

$$\langle\gamma\rangle \leq \frac{N(N-1)A^2}{nL^3} \int_0^L\!\!\int_0^L\!\!\int_0^L f_2(\tilde{x}_1,\tilde{x}_2)\phi_x(x,\tilde{x}_1)\phi_x(x,\tilde{x}_2)d\tilde{x}_1 d\tilde{x}_2 dx$$

$$(2.30) \quad + \frac{NA^2}{nL^2}\int_0^L\!\!\int_0^L f_1(\tilde{x}_1)\phi_x^2(x,\tilde{x}_1)d\tilde{x}_1 dx + \frac{1}{L}\int_0^L f_1(\tilde{x}_1)(1+A\phi(\tilde{x}_1,\tilde{x}_1))^2 dx_1$$

$$+ \frac{N(N-1)}{NL^3}\cdot\int_0^L\!\!\int_0^L f_2(\tilde{x}_1,\tilde{x}_2)(1+A\phi(\tilde{x}_1,\tilde{x}_1))A\phi(\tilde{x}_2,\tilde{x}_1)d\tilde{x}_1 d\tilde{x}_2$$

$$+ A^2\frac{N(N-1)}{nL^3}\int_0^L\!\!\int_0^L f_2(\tilde{x}_1,\tilde{x}_2)\phi^2(\tilde{x}_2,\tilde{x}_1)d\tilde{x}_1 d\tilde{x}_2$$

$$+ \frac{A^2 N(N-1)(N-2)}{nL^4}\int_0^L\!\!\int_0^L\!\!\int_0^L f_3(\tilde{x}_1,\tilde{x}_2,\tilde{x}_3)\phi(\tilde{x}_3,\tilde{x}_1)\phi(\tilde{x}_2,\tilde{x}_1)d\tilde{x}_1 d\tilde{x}_2 d\tilde{x}_3 .$$

In the infinite limit $\phi(x,\tilde{x}_i)$ may be replaced by $e^{-\mu|x-\tilde{x}_i|}$ in (2.30), yielding an upper bound involving f_1, f_2, f_3, A, and μ. This expression can then be minimized with respect to μ and A to obtain an optimal bound involving f_1, f_2, and f_3. For the uniform distribution we have

$$(2.31) \qquad \gamma \leq A^2\mu + (1+A)^2 + \frac{A^2 n}{\mu} .$$

and the optimization yields

$$(2.32) \qquad \gamma \leq (1 + \tfrac{1}{2}\sqrt{n})^{-1} = 1 - \frac{1}{2\sqrt{n}} + \frac{1}{4n} + \ldots .$$

The first two terms on the right of (2.32) are equal to the leading terms in the expansion of the Brinkman value (2.12) for large n, while the 1/n terms are not the same. Also, the bound (2.32) is worse than the previous upper bound if $n < 1$. Thus the sharper result for large n (assuming the accuracy there of the Brinkman value) is obtained at the expense of estimates at small n.

A Case of Second-order Brinkmanship. The "first-order" Brinkman approximation (2.14) closes the hierarchy of averaged equations with (2.13). We suppose now that (2.13) is retained and that the solution of (2.13), if the right-hand side is known in terms of $<y>_1$, would be of the form

$$(2.33) \qquad <y>_1 = <y> + <y(\tilde{x}_1)>_1 G(x-\tilde{x}_1) , \quad G(-x) = G(x) , \quad G(0) = G_0 .$$

This ansatz will be used at the next level to obtain a second-order closure. Averaging with two points fixed yields

$$(2.34) \qquad <y(x)>_{12} = -1 + <y(\tilde{x}_1)>_{12}\delta(x-\tilde{x}_1) + <y(\tilde{x}_2)>_{12}\delta(x-\tilde{x}_2)$$
$$+ n<y(\tilde{x}_3)>_{123}\Big|_{\tilde{x}_3 = x} .$$

We assume that the solution of (2.34) is well approximated in terms of G by the expression

$$(2.35) \qquad <y(x)>_{12} \simeq <y> + <y(\tilde{x}_1)>_{12} G(x-\tilde{x}_1) + <y(\tilde{x}_1)>_{12} G(x-\tilde{x}_2) .$$

Noting (from (2.33)) that

$$(2.36) \qquad 1 - G_0 = 1/\gamma ,$$

we may solve (2.35) for $<y(x_2)>_{12}$ to obtain

$$\tag{2.37} <y(\tilde{x}_2)>_{12} = \frac{\gamma <y>}{1-\gamma G(\tilde{x}_2-\tilde{x}_1)}.$$

Using the last expression in (2.13), we see that G now satisfies the nonlinear equation

$$\tag{2.38} G''(x) = \delta(x) + \frac{G n \gamma}{1-\gamma G}.$$

With the conditions $G(\infty) = 0, G'(0+) = 1/2$, we find upon integration of (2.38) that

$$\tag{2.39} \tfrac{1}{2}(G'(0))^2 = \tfrac{1}{8} = -nG_0 - \tfrac{n}{\gamma} \ln(1-G_0)$$

yielding

$$\tag{2.40} n^{-1} = \tfrac{8}{\gamma}(1 - \gamma - \ln(2-\gamma)),$$

a second-order Brinkman approximation to $\gamma(n)$. It is easy to check that if the series for the logarithm is approximated by

$$\tag{2.41} \ln(2-\gamma) \simeq 1 - \gamma - \tfrac{1}{2}(1-\gamma)^2,$$

then we recover the first-order result (2.12).

Summary of estimates in the model problem. We show in Figure 2 the various bounds and approximations to γ obtained above, all for the completely uniform distribution, and compared with Monte Carlo simulation with 200 points and 100 trials (see Appendix). The standard deviations obtained in the simulations are shown in Table 1.

n	.5	1	2	4	8	16	32	64	128
γ	.436	.563	.676	.766	.835	.885	.921	.946	.964
Standard Deviation	.047	.047	.042	.035	.029	.021	.016	.013	.010

Table 1. Monte Carlo simulation in the model Problem (2.2) for a Poisson distribution of d_i, using 200 points x_i and 100 trials.

Figure 2. Estimates for γ with a uniform distribution of springs for the model problem (2.2). (a), (b) are the upper and lower bounds given by (2.25). (c) is the upper bound (2.32). (d) is the first-order Brinkman estimate (2.12). (e) is the second-order result (2.40).

The following observations can be made: (a) The bounds (2.25) do not jointly imply a good approximation to γ except for small n, where they converge to the asymptotic expression $\gamma \sim 2n$. (b) First-order Brinkman theory actually exceeds the upper bound (2.25) for $n < .5$. (c) The second-order value is remarkably close to the simulations down to $n = .5$.

We remark that no lower bound to go with (2.32), which might yield a proof of the asymptotic validity of the Brinkman approximation for large n, has been given here, although a suitable choice of trial function g in (2.21) may establish such a result. Also it is not immediately obvious that a useful third-order theory can be found in a hierarchy of Brinkman approximations, since the one-point function G satisfies a nonlinear equation, indicating that the effective medium is not, at that level of approximation, a linear elastic bed. Of course such "higher-order" extensions are not uniquely defined and there is perhaps a procedure different from the one used here which would clarify, at least in principle, how an "nth order" Brinkman approximation should be defined.

3. Attachments at Finite Intervals.

The condition that $y(x)$ be determined as above through a set of point attachments at x_i is unrealistic when modeling finite porosity, so we consider now the Brinkman approximation when the one-dimensional model is modified as shown in Figure 3. Each spring is now to be attached at the center of a horizontal rigid element of length d. It is assumed that the pressure difference across the membrane also acts on these rigid elements (since this mimics the Archimedean force experienced by a finite body in a permeable bed when there is a pressure gradient through the bed).

Figure 3. Spring attachment to a rigid element.

Monte Carlo simulation with $d > 0$ may in fact be used to obtain data for $d = 0$ by the following transformation: If $\gamma_0(n_0)$ corresponds to $d = 0$, then

$$\langle y_1 \rangle_1 = \frac{1}{n_0} + d = \frac{1+n_0 d}{n_0} = \frac{1}{n} ,$$

(3.1)

$$\langle y \rangle = d + \frac{dn}{n_0} + \frac{n}{n_0^2}$$

as can be seen from the definitions of the averages applied to the collection of membrane and rigid elements. Thus, using the values in Table 1 for n_0 and γ_0,

(3.2) $$\gamma = \frac{(1+n_0 d)^2 \gamma_0}{1+\gamma_0 n_0 d + \gamma_0 n_0 d(1+\gamma_0 n_0 d)}$$

will yield Monte Carlo values for a Poisson distribution of gaps with density $n_0 = n/(1 - nd)$.

The simplest second-order Brinkman approximation which might be applicable here assumes the two-point distribution function f_2 to have values 0 or 1. If the test spring attaches to the elastic bed after a free gap of width a, then $G(x)$ satisfies

(3.3) $\quad G'' = -n ,\qquad\qquad\qquad 0 < x < a$

$\qquad\quad = \frac{n\,G}{(1-G)} ,\qquad\qquad\quad x > a ,$

$$G(0) = \frac{\gamma-1}{\gamma}, \qquad G'(0) = \frac{1}{2}(1-nd),$$

G and G' continuous at $x = a$.

Thus

(3.4) $\qquad G(a) = \frac{\gamma-1}{\gamma} + (1-nd)\frac{a}{2} - \frac{a^2 n}{2},$

$$\frac{1}{8}[1-n(d+2a)]^2 + nG(a) + \frac{n}{\gamma}\ln(1-\gamma G(a)) = 0.$$

In this case it is seen that γ becomes unity, corresponding to "blocking," at a finite value of n.

In Table 2 we give some values obtained from (3.4), compared with the computations. Note that the choice $a = d/2$ tends to overestimate

	d = .1			d = .5		
n	a = 0	.05	mc	a = 0	.25	mc
.5	.458	.470	.460	.490	.562	.547
1.0	.588	.612	.609	.638	.770	.743
2.0	.704	.746	.750	.765	.970	.892
4.0	.799	.865	.865			
8.0	.872	.965	.943			

Table 2. Comparison of Monte Carlo (mc) values with those obtained from (3.4).

γ. This suggests that the correct 2-point distribution function, corresponding to independent allocation of segments with no overlap, will improve the agreement. In one dimension this distribution function can be described explicitly [9]. With this modification we have, in place of (3.3),

(3.5) $\qquad G'' = g(x - d/2)L(G), \qquad L(G) = \gamma G/(1-\gamma G),$

where

(3.6) $\qquad g_d(x) = \sum_{k=0}^{\infty} \frac{n_0^{k+1}}{k!} H(x-kd) e^{-n_0(x-kd)} (x-kd)^k,$

and $H(x)$ is the Heaviside function. We also are led to pose the following question: Is there a choice of $L(G)$ in (3.5) which is in some sense optimal in yielding values of γ which obey or nearly obey (3.2)? We have not as yet solved (3.5), (3.6) for $\gamma(n)$, but such a computation would be useful in conjunction with any systematic study of a hierarchy of Brinkman approximations.

4. The Role of Dimension

The one-dimensional case is formally analogous to a class of higher-dimensional problems involving disordered media, although there are implications of the dimension, and of the specific operators involved, when invoking Brinkman's approximation. As Batchelor [10] has noted, a large number of elliptic problems involving Dirichlet or Neumann data at random boundaries are formally equivalent, and we shall restrict attention here to the permeability problem for slow viscous flow.

Indeed, (2.2) is equivalent to a rather artificial problem of this kind. If $y(x)$ is interpreted as a velocity of a viscous fluid in a direction orthogonal to x the x-axis, and if the planes $x = x_i$ are imagined to be porous sheets, then the velocity $y(x_i)$ will be nonzero at the sheets owing to the stress, and it is reasonable to postulate that the jump in y' will be proportional to the slip velocity. Then (2.2) is obtained and γ is equivalent to an inverse permeability. In addition the constant -1 on the right of (2.2) may be identified with the pressure gradient needed to sustain the flow.

Two dimensions. Since Monte Carlo simulation should be feasible for some two-dimensional problems, the following example is of interest: consider parallel viscous flow orthogonal to the x-y plane, in the domain exterior to randomly placed, nonoverlapping circles of diameter d. If $w(x,y)$ denotes the velocity, and g is the pressure gradient in the z direction, then when the fluid adheres to the boundaries we have

$$(4.1) \qquad \mu \nabla^2 w = g, \qquad w = 0 \quad \text{on} \quad C_i, \qquad i = 1, 2, \cdots.$$

where μ is the fluid viscosity and C_i denotes the ith boundary. Then

$$(4.2) \qquad F_i = \int_{C_i} \frac{\partial w}{\partial n} \, ds$$

is the force per unit length down the ith cylinder, and the problem is to determine

$$(4.3) \qquad \gamma = \frac{<F_i>}{\mu <w>}.$$

This problem has been treated by Howells using Brinkman's approximation [11]. The equivalent diffusion problem has recently been analyzed in considerable detail by Talbot and Willis [12], who give upper and lower bounds on γ analogous to those of Section 2.

Three dimensions. Viscous flow past a random bed of non-overlapping spheres has been discussed by Tam [13], Childress [14], Howells [11], Saffman [15], and Hinch [16]. For small values of the volume density c of the array, and for spheres of fixed diameter d, the expected force F, divided by the force F_0 of a single sphere in the same mean flow, is found to have an expansion

$$(4.4) \quad \gamma = 1 + \frac{3}{\sqrt{2}} c^{\frac{1}{2}} + \frac{135}{64} c \ln c + ch(f_2) + O(c^{3/2} \ln c),$$

where we indicate the first dependence of the two-point distribution function for the centers of the spheres. For an f_2 which is zero for overlap, one otherwise, there results $h \sim 16.5$. Unfortunately, the expansion (4.4) is less useful than the Brinkman value for c of practical interest. Childress [14] proposed the second-order extension of Brinkman's approximation used in Sections 2 and 3 above, in the context of Stokes's equations and point particles, and noted that it then contains the terms in (4.4) not specifically due to the finite size of spheres. However no application was made to finite c.

It should also be noted that an equivalent "second-order" scattering theory in three dimensions was proposed for the coherent field by Twersky [17].

Our reason for listing these problems here is to point out differences with the one-dimensional case. In the first place the relevant fundamental solutions for the case of point "scatterers" are singular in the higher dimensions. Thus, in the three-dimensional point particle problem, the terms $y(x_i)$ in (2.2) are replaced by regular parts evaluated at the centers. (The two-dimensional problem is special owing to the logarithmic behavior of the fundamental solution.)

In the second place, there is the essential difference in the expected range of validity of Brinkman's approximation. In one dimension, the expected interparticle distance is $O(n^{-1})$ while the "shielding" distance in Brinkman's theory is $O((\gamma n)^{-\frac{1}{2}})$. Thus, as noted previously, the continuous bed is reasonable when $n \gg 1$, where in fact $\gamma \sim 1$. In dimension N the first of the above estimates becomes $n^{-1/N}$. Thus for $N = 2$ there is no preferred range of n, except as provided by the value of γ. In three dimensions, by the same reasoning, Brinkman's result should apply to small n, or more exactly to small volume density c; indeed, the first two terms on the right of (4.4) are obtained by expanding the solution of (1.1) for small c [13,14].

Nevertheless, the point seems to be that the only reason for using these approximations is to obtain useful expressions outside their expected range of validity, and to test these ideas we need more data over a range of values of volume fraction, the important asymptotics being in the limit of dense packing.

The author would like to thank George Papanicolaou, Jerry Percus, and Charles Peskin for helpful discussions.

REFERENCES

1. Brinkman, H.C., Appl. Sci. Res. A1, 27, 1947.
2. Foldy, F.L., Phys. Rev. 67, 107, 1945.
3. Twersky, V., J. Res. Nat. Bur. Standards 64D, 715, 1960.
4. Lax, M., in "Stochastic Differential Equations," SIAM-AMS Proceedings, Vol. VI, 1973.
5. Happel, J. and Brenner, H., Low Reynolds Number Hydrodynamics, Prentice-Hall, 1965.
6. Karlin, S., A First Course in Stochastic Processes, Academic Press, 1966.
7. Bazer, J., J. Soc. Indus. Appl. Math. 12, 539, 1964.
8. Frisch, H.L. and Lloyd, S.P., in Mathematical Physics in One Dimension, Academic Press, 1966.
9. Percus, J.K., in Studies in Statistical Mechanics, edited by Lebowitz and Montroll, North Holland, Amsterdam, 1981.
10. Batchelor, G.K., An. Rev. Fluid Mech. 6, 227, 1974.
11. Howells, I.D., J. Fluid Mech. 64, 449, 1974.
12. Talbot, D.R.S. and Willis, J.R., Proc. Roy. Soc. Lond. A, 370, 351, 1980.
13. Tam, C.K.W., J. Fluid Mech. 64, 449, 1974.
14. Childress, S., J. Chem. Phys. 56, 2527, 1972.
15. Saffman, P.G., Stud. Appl. Math. 52, 115, 1973.
16. Hinch, E.J., J. Fluid Mech. 83, 695, 1977.

APPENDIX

In the interval $x_{i-1} \leq x \leq x_i$ we have

(A.1) $$y = -\frac{1}{2}(x-x_{i-1})^2 + a_i(x-x_{i-1}) + y_{i-1},$$

so that

(A.2) $$y_1 = y_{i-1} + a_i d_i - \frac{1}{2}d_i^2, \qquad d_i = x_i - x_{i-1},$$

(A.3) $$a_i = a_{i-1} + y_i - d_{i-1}.$$

We combine the last two equations to obtain

(A.4) $$d_i^{-1} y_i + \left(1 - d_i^{-1} - d_{i-1}^{-1}\right) y_{i-1} + d_{i-1}^{-1} y_{i-2} + \frac{1}{2}(d_i + d_{i-1}) = 0,$$

$$d_0 = d_N, \quad y_0 = y_N.$$

The Monte Carlo simulation involves, for each trial, the inversion of a tridiagonal matrix plus two corner entries to take care of the periodicity condition. Two tridiagonal inversions are thus needed. The value of γ is then given by

(A.5) $$\gamma = \left(\sum_{i=1}^{N} g_i\right) \left(\sum_{i=1}^{N} d_i^3/12 + (y_i + y_{i-1}) d_i/2\right)^{-1}.$$

TOPOLOGY, GEOMETRY, AND PHYSICAL PROPERTIES OF POROUS ROCKS

Morrel H. Cohen
James Franck Institute and Dept. of Physics
The University of Chicago
Chicago, IL. 60637

AND

Charlotte Lin
Schlumberger-Doll Research Center
P.O.Box 307
Ridgefield, CT. 06877

ABSTRACT

By use of the concepts of skeletization and deformation retract, we characterize the geometry and topology of a porous sedimentary rock in particularly simple ways. We briefly introduce the underlying topological concepts and present a skeletization procedure which leads to clear definitions of such concepts as grain, contact, pore chamber, channel, and throat. We apply this procedure in developing novel formulations of the problems of nuclear magnetic relaxation within the pore space, of steady flow through the pore space, and of the frame moduli. We show how the ambiguity between pore chambers and channels can be exploited for the NMR and flow problems. In particular, we find that a flow problem can be reduced to a resistance network problem, but the network is not a deformation retract of the pore space. The frame moduli problem can be mapped into the long wavelength the limit of a random "lattice" - vibration problem.

I. INTRODUCTION

In the present state of rock physics one can formulate an immense number of significant, unanswered questions about the physical properties of detrital sedimentary rocks. For example, what does the magnetic relaxation of protons in the pore water of such rocks tell about the pore space geometry? Why does the self-similar model account so well for the dc electrical conductivity associated with electrolyte in the pore space?[2,3,4] How could one obtain comparable accuracy in calculating the frame modulus in the Biot theory?[5] How could one calculate the permeability for flow through the pore space[6] to such accuracy? What would the relation of the permeability and the conductivity be in such calculations?[6] The solutions to these illustrative problems all require explicit quantitative knowledge of the geometry and topology of the porous rocks.

Consider the question of the accuracy of the selfsimilar model. It involves two sets of approximations, one set approximating the geometry and topology and one set introduced in calculating electrical conductivity for a given geometry. Without deeper study of both, one cannot understand why the model should work. The calculation procedure used is the effective medium approximation, the convergence of which is relatively easy to study. What remains is the accuracy of the description of the geometry and the topology within the selfsimilar model. Addressing that question entails learning more about the geometry and topology.

A very powerful means of doing so is through reconstructive modelling from serial sections using quantitative characterization of the topology and geometry. We describe procedures for this elsewhere.[7] Here we briefly summarize the required topology and a simple procedure for modelling the grain and pore spaces which leads to sharp definitions of grains, contacts, pores, channels, and throats. We next illustrate how these ideas can be useful in solving various problems which arise in the study of the physical properties of sedimentary rocks.

II. TOPOLOGY

Consider the simplest type of sedimentary rock, a clean sandstone. An ideally clean sandstone is a two-phase material composed of connected rock grains and the associated intergranular pore space. Each grain can be represented by a point (<u>node vertex</u>) within it. Each contact between two grains can be represented by an arc (<u>branch</u>, <u>edge</u>) connecting their two representative points. The <u>graph</u> comprised of the vertices and edges partitions any surface on which it lies into <u>faces</u>. The faces partition the three-dimensional space occupied by the sandstone into cells which fill the space. This construct of vertices, edges, faces, and cells is a <u>complex</u> which represents the grain space, G. G may also be represented by a standard topological solid, X, such as a solid <u>torus</u> (doughnut), an <u>n-torus</u> (doughnut produced with an N-holed doughnut cutter), an n-torus with internal <u>holes</u> which are themselves m-tori, etc. X^C is all of three-dimensional space not in <u>X</u>, including the m-torus holes in X. The analogous representation for the pore-space P has vertices in the pore chambers and edges passing through the channels between them. Note that X can represent either G or P and X^C either P or G, respectively and that $G^C = P$.

The graph introduced above is a <u>deformation retract</u> D_G of G. A subset D of a set X is a deformation retract of X if, given the function f(x) mapping X onto D, there is a function $g(x,t)$ which is continuous in x and t such that

$$g(x,t) = x, \quad \text{if } t = 0$$
$$= f(x), \quad \text{if } t = 1$$
$$= \text{some element of } X \text{ for all } t \text{ in the interval } [1,0]$$

That is, g shrinks X continuously in space and time onto D, which is a <u>skeleton</u> of X. In the next section we discuss specific methods of generating deformation retracts D_G of P. In reconstructive modelling of G and P, it is far easier to specify D_G and D_P than G and P themselves.

We use the <u>Betti numbers</u> to characterize the grain and pore spaces topologically. The definition of the Betti numbers requires a nontrivial amount of topology which is summarized by Barret and Yust.[8] "The nth <u>Betti number</u> of a complex . . . (is) . . . the maximum number of homology-independent N-cycles".[8] The Betti numbers β_i of X are the same as those of any deformation retract D of X. It is easier to determine the β_i on D than on X.

It is enough for our purposes to know what $\beta_0, ---, \beta_3$ tell us about X. The <u>zeroth Betti number</u> β_0 of X gives the number of separate components comprising X. The grain space G of a well-consolidated sandstone has only one component, $\beta_0(G) = 1$, unless there are N loose grains or grain clusters inside when $\beta_0(G) = N + 1$. Similarly, $\beta_0(P) > 1$ can indicate the presence of porosity which does not contribute to permeability and which would be measured by logging techniques such as hydrogen detection but missed by a porosimeter measurement. An instance would be an isolated ring formed by dissolution of a grain or grains, followed by encapsulating cementation or quartz overgrowth. The Alexander duality theorem states that the <u>first Betti number</u> or cyclomatic number $\beta_1(X)$ of X equals that of X^c, $\beta_1(X) = \beta_1(X^c)$. Determining $\beta_1(G)$ determines $\beta_1(P)$ as well. β_1 equals $E - V + 1$, where E is the number of edges and V the number of vertices in our graph D_G. It measures the number of "circuits" or closed loops in the graph, and by the Alexander duality theorem it suffices to determine either $\beta_1(D_G)$ or $\beta_1(D_P)$ if the representations are true deformation retracts. When D_G is large, β_1 can be approximated by $E - V$ so that

$$C = \beta_1/V = E/V - 1$$

Defining Z as the average coordination number of the graph, we have

$$Z = 2E/V = 2(1 + C)$$

For the <u>second Betti number</u> $\beta_2(X)$, we have the relation $\beta_2(X) = \beta_0(X^c) - 1$ so that

$\beta_2(G) = \beta_0(P) - 1$. Since $\beta_0(P)$ is the number of separate pore subsystems, $\beta_2(G)$ gives the number of internal surfaces. The third Betti number $\beta_3(X)$ is zero for the three-dimensional spaces we expect to encounter.

The Euler-Poincare formula equates the alternating sum of the first four Betti numbers to the Euler characteristic X, which is the alternating sum of the numbers of vertices, edges, faces, and cells in our complex. The latter numbers and the Betti numbers are topological invariants; any distortion of the net without the cutting of edges and vertices leaves them unchanged. The Euler characteristic is therefore a topological invariant; it is the same for complexes with the same "connectivity". Thus the connectivity of G, say, can be characterized by careful deformation to a graph D_G, by determination of the Euler characteristic of D_G which classifies it, and by subsequent mapping to an appropriate solid X with Betti numbers which give more detailed information about the components, coordination number, and internal surfaces of G.

The notion of the genus of a surface is sometimes substituted for the first Betti number of the complex containing the graph lying on the surface. While it is a technical misstatement to say that the genus and β_1 are identical, their two values are equal for graphs lying on surfaces in complexes. Careful specification of the graph at the exterior boundary of the sample will insure that the graph lies on a surface without boundary as is required for surfaces in complexes, and it is a theorem that topologically the only such surfaces are spheres, tori, and n-hole tori. Rhines[9] has used genus normalized to unit volume. One can also use genus per vertex which equals C defined above and relates directly to the coordination number of the graph.

III. SKELETIZATION AND MODELLING

The preceding section shows the convenience and utility of characterizing the topology and geometry of the grain and pore spaces through their deformation retracts or skeleton graphs D_G and D_P. In the present section we describe a particular way to construct these deformation retracts, that is, a particular skeletization,[10-15] which leads, as mentioned in the introduction, to sharp definitions of grains, contacts, pore chambers, channels, and throats.

Center on each point of the grain-pore interface an arbitrarily small sphere of radius ε. Erode the grain-space G by deleting its intersection with the collection of spheres, producing a new grain-pore interface and opening the pore space. Iterating the process ultimately causes G to disappear. The intermediate effects are to smooth small-scale roughness in the interface and, more importantly, progressively to disconnect G at progressively larger local volumetric minima.[16,17] A region of grain space which becomes isolated by these disconnections is either a grain or a cluster of grains in contact which subsequently disconnect. When an isolated grain has become convex, its centroid can be taken as the grain center.

Return to those instants during the erosion process at which manifest separation appears between two regions of G previously in contact. Run the erosion process backwards and reestablish contact between the regions. Each point of first contact represents the "center" of an intergranular contact. The minimum-area cross section which contains the contact center can be taken as the contact surface between the two regions at any stage of erosion prior to disconnection. The grain space is now partitioned into individual grains bounded by the contact surfaces and the pore-grain interface.

Return now to the original G. Embedded within G is a set of grain centers and contact surfaces. To represent G by a network take each grain center as a vertex. Connect them by edges which are arcs through each contact surface. This skeleton of G is a deformation retract D_G because it can be obtained from G by reversing an opening process which utilize spheres of arbitrarily small radius centered on each point of D_G and iterates on the resulting surface until each point of the original grain-pore interface is reached.

Continue the reversed erosion of G beyond the original pore-grain interface, thus eroding P. Having identified individual grains, we can recognize the point-by-point instances of one grain's pore-grain interface growing into contact with that of another grain. The collection of interface contact surfaces formed by these events upon P's disappearance is a deformation retract of P, D_p. The interface contact surfaces are continuations of the grain contact surfaces. The total collection of surfaces thus partitions 3-space into cells, each enclosing an individual grain. Deforming each cell surface into a topologically equivalent polygonal surface produces polyhedral cells which are topologically equivalent to Voronoi polyhedra.

The notions of pore "chambers", "channels", and "throats" can now be developed. The partitioning of space into cells by the contact surfaces defines a network of vertices and edges which is also a deformation retract of P, D_p. Edges and vertices then are the line segments and points, respectively, formed by the intersections of three or more finite contact surfaces. The point intersections occur in the pore chambers; that is, the pore chambers are the spaces contiguous to the vertices of D_p. The line intersections occur in the channels connecting pore chambers; that is, the channels are the spaces contiguous to the edges of D_p not already included in the pore chambers. Throats are the locally minimal area cross-section loci in the channels, and their locations along the edges delineating the channels are particular to the local geometry. The geometric features specified by these definitions are clearly and simply displayed by sphere packs. The construction of D_G and D_P and the definitions of grains, contacts, pore chambers, channels, and throats can be used as a basis for modelling the pore space in different ways appropriate to the study of different physical properties. Examples are given in the next section.

IV. MODELLING THE GRAIN AND PORE SPACES.

In this section the utility of the ideas in the two previous sections is illustrated by accurately modelling the pore and grain spaces as appropriate to the calculation of several different physical properties of sedimentary rocks.

A. Nuclear magnetic relaxation.

Cohen and Mendelson[1] have shown that the nuclear magnetization M in the direction of a magnetic field H which is associated with a density N_b of protons in the pore water obeys the following equations

$$\frac{\partial M}{\partial t} = D \nabla^2 M - \frac{M - M_\infty}{T_{1b}} \tag{1}$$

$$\frac{\partial M(s)}{\partial t} = - \frac{D}{\ell} \hat{n} \cdot \vec{\nabla} M(s) - \frac{M(s) - M_\infty}{T_{1s}} \tag{2}$$

$$M = M_0 \text{ at } t=0 \tag{3}$$

In Eq. (1), $M_\infty = \chi H$, where χ is the bulk susceptibility of protons in water D is the proton diffusion coefficient in water, and T_{1b} is the bulk relaxation time (of order sec). In Eq.(2), M(s) is the limiting value of M as a point on the rock-water interface is approached from within the water. ℓ is the ratio N_s/N_b of the surface density N_s of protons on the hydrated rock surface to the bulk proton density N_b (of order an Angstrom unit). $\hat{n}$ is the unit normal vector pointing outward from the fluid. T_{1s} is the relaxation time of protons at the rock-water interface (of order 10^{-4} sec).

We now take advantage of the ambiguity that exists in the partitioning of the pore space into pore chambers and channels described in § III. No boundaries have been placed between the chambers and the channels. In the present instance it is convenient to eliminate the channels completely, dividing all of the pore space into separate pore chambers, or simply pores, by passing a surface through the throat of each channel. Each pore i, i = 1 ---N, has volume Vi and surface S_i. S_i consists of a portion of the rock-water interface, S_i, and the boundary surfaces through the throats separating i from its neighbors j, S_{ij}

$$S_i = S_i^r + \sum_j S_{ij} \tag{4}$$

Cohen and Mendelson show that under the conditions which prevail in the rocks of interest, the magnetization is nearly uniform within each individual pore. This permits simplification of Eq.(1) and (2) to

$$\frac{dM_i}{dt} = -\frac{M_i - M_\infty}{T_i} + \frac{1}{V_i} \sum_j K_{ij} (M_j - M_i) \tag{5}$$

In (5) M_i is the mean magnetization in pore i, and T_i is the relaxation time for M_i when coupling to the neighboring pores can be ignored,

$$\frac{1}{T_i} = \frac{1}{T_{1b}} + \frac{\ell S_i^r}{V_i} \left(\frac{1}{T_{1s}} - \frac{1}{T_{1b}} \right) \tag{6}$$

The sum on j is over all contiguous pores j, and the coupling coefficients are given by

$$K_{ij} = K_{ji} = D\, S_{ij} / L_{ij}, \tag{7}$$

where the length L_{ij} is defined implicitly by the solution to (1) - (3). In the cases of interest, the throat dimensions are small compared to the channel dimensions and L_{ij} is approximately the throat diameter.

Equation (5) defines a problem in which there is a quantity M_i associated with each vertex i of the deformation retract D_p. The time dependence of the M_i is governed by a matrix with diagonal elements T_i^{-1} associated with each vertex and off-diagonal elements $V_i^{-1} K_{ij}$ associated with the directed edge ij of D_p. The values of the T_i^{-1} are random, as are the values of the $V_i^{-1} K_{ij}$. The connectivity of the matrix is also random; that is, the network D_p is topologically disordered.

Analysis of (5) is greatly facilitated by the fact that it has the form of equations extensively studied in the microscopic theory of disordered materials. Drawing upon that experience Cohen and Mendelson have shown the following:

1. In the limit that the $V_i^{-1} K_{ij}$ are negligible compared to the variation of the T_i^{-1}, the relaxation function of the magnetization is the Laplace transform of the probability distribution of the T_i^{-1}. According to (6) the probability distribution of the pore surface to volume ratios S_i^r / V_i can then be determined up to a scale factor.

2. In the limit that the $V_i^{-1} K_{ij}$ dominate the variation of the T_i^{-1}, the relaxation of the magnetization is exponential, with a T_i which contains the surface to volume ratio of the whole pore space.

3. In the intermediate case, the nuclear magnetic relaxation depends on the distributions of the S_i^r/V_i and the S_{ij}/L_{ij} and on the topology of D_p. Disentangling these dependences can be carried out with the aid of the microscopic theory of disordered materials.

B. Steady flow problems

The steady flow of fluid through a porous rock in response to a pressure head[18] or the steady flow of electricity through a conducting fluid within the pore space[4] is more conveniently addressed with the opposite resolution of the pore-chamber/pore channel ambiguity. Consider the case of the steady flow of electric current. The equations governing the flow are

$$\nabla^2 \psi = 0 \tag{8}$$

$$\hat{N} \cdot \vec{\nabla} \psi (s) = 0 \tag{9}$$

$$\vec{E} = -\vec{\nabla} \psi \tag{10}$$

$$\vec{j} = \sigma_w \vec{E} \tag{11}$$

with appropriate boundary conditions on the electrostatic potential ψ at the outer boundaries of the rock. Here $\vec{E}$ is the internal electric field, $\vec{j}$ the current density, and σ_w the conductivity of the electrolyte. Suppose that Eqs.(8) and (9) are solved subject to the external boundary conditions so that ψ, $\vec{E}$, and $\vec{j}$ are known everywhere within the pore fluid.

Consider now a vertex i of D_p at which r edges intersect. There will be r channels entering the corresponding pore chamber. Suppose that current flows into the pore chamber through p of the r channels and out through q of them, $p + q = r$. The current flow is normal to the surfaces of constant electrostatic potential, which decreases in the direction of flow. Above some value ψ_{ni} of ψ, there will be p distinct pieces of each surface of constant ψ within the p channels with inward flowing current, and below some other value ψ_{1i} of ψ, there will be q distinct pieces within the q channels with outward flowing current. Let $\alpha_i = (m_i, n_i)$ specify the number of pieces of the surfaces of constant ψ near the vertex i for ψ in $[\psi_{ui}, \psi_{\ell i}]$, m_i being the number across which there is inward flow and n_i that for outward flow. α_i changes from $(p,0)$ at ψ_{ui} to $(0,q)$ at $\psi_{\ell i}$ in a number of steps larger than one in general. At each such change, $\psi = \psi_{\sigma i, \sigma} = 1, --- \nu_i$ ($\psi_{1i} = \psi_{ui}, \psi_{\nu_i i} = \psi_{\ell i}$) defines a critical surface separating a region of pore space with $\alpha_i = \alpha^\sigma_i$ from one with $\alpha_i = \alpha_i^{\sigma+1}$. We can use these critical surfaces to partition the pore chamber i. This can be repeated for all pore chambers until the entire pore space is so partitioned.

The pore space is thus divided entirely into channels by successive critical surfaces along the lines of current flow. Each critical surface defines a vertex, and each channel defines an edge connecting it to another vertex. The network F so constructed, however, is not topologically equivalent to D_p and therefore does not represent P. For each vertex i in D_p there are ν_i vertices in F. F reduces identically to D_p only in the improbable event that all ν_i are unity. However, if most of the voltage drops $\psi_{ui} - \psi_{\ell i}$ are small compared to the differences in the means ($\psi_{\sigma i} - \psi_{\sigma j}$) between adjacent vertices in D_p, one can ignore the fine structure associated with each vertex, lump the channels between ψ_{ui} and $\psi_{\ell i}$ with those leading into ψ_{ui} or out of $\psi_{\ell i}$ for convenience and arrive at a network equivalent to D_p.

The current flow within each channel as defined above is governed by the boundary conditions $\hat{n} \cdot \vec{j} = 0$ at the rock fluid interfaces and $\hat{n} \times \vec{j} = 0$ at the critical surface of constant voltage. We can thus associate with each vertex i of F (rigorously or of D_p approximately) a voltage ψ_i and with each edge ij a current

$$I_{ij} = \int j ds, \tag{12}$$

where the integration is over the area of any surface of constant voltage within the channel ij. Ohm's law holds for the channel and may be written in the form

$$I_{ij} = G_{ij} (\psi_i - \psi_j), \tag{13}$$

where the conductance G_{ij} is given by

$$G_{ij} = \sigma_w \int \vec{E} \cdot d\vec{s} / \int_j^i \vec{E} \cdot d\vec{r} \tag{14}$$

Consideration of the geometry of the channel can yield rough estimates of G_{ij}, for example $G \simeq \sigma_w A/L$ where A is an appropriate mean channel crossection and L an appropriate mean channel length. Integrating $\vec{\nabla} \cdot \vec{j} = 0$ within a volume within the pore chamber which encloses the critical surface yields Kirchoff's law

$$\sum_j I_j = \sum_j G_{ij} (\psi_i - \psi_j) = 0 \tag{15}$$

We thus have a random resistance network problem with a voltage ψ_i at each vertex i and a resistance G_{ij}^{-1} along each ij of the network D_p.

Having made contact with random resistance networks, we have at our disposal a substantial body of work which we can carry over to the problem of the dc conductivity of porous rocks. One can proceed with the analysis of fluid flow problems on the same basis, establishing a set of Kirchoff equations for the above network F containing the permeabilities of the individual channels K_{ij} and then going on to derive the permeability of the rock.

C. The frame moduli.

Turning now to the grain space, a challenging problem is the calculation of the frame moduli, the elastic moduli of the rock in the absence of a pore fluid. When the grain contact areas are small relative to the grain size, stress concentration is such that elastic distortion occurs primarily in the vicinity of the contacts. Neglecting overlap of the displacement fields associated with different contacts on the same grain, and assuming sound wavelengths much larger than the grain size, one can derive straightforwardly a set of equations of motion for each grain:

$$M_i \frac{d^2 \vec{u}_i}{dt^2} = \sum_j \overleftrightarrow{K}_{ij} \cdot (\vec{u}_j - \vec{u}_i) \tag{16}$$

Here m_i is the mass of grain i and $\vec{u}_i$ its displacement, and K_{ij} is the force-constant tensor coupling grain i to grain j across the contact area. Evaluation of K_{ij} requires a detailed study of the elastic properties of the contact and requires specification of its geometric details. However, estimations are readily made. We have thus mapped the problem of wave propagation in and the elastic properties of a complex continuum into the same problem for the random network D_G with masses M_i associated with the vertices and force constants $\overleftrightarrow{K}_{ij}$ associated with the edges. The problem has been studied in detail in the microscopic context of phonons in disordered materials. The normal mode spectrum, sound velocities, and the frame modulii can thus be studied with the aid of an extensive literature. However, the dispersion of the sound velocity is incorrectly given in this theory, as we show in a more detailed study.

V. CONCLUSIONS AND DISCUSSION

We have provided detailed, particular answers to such questions as what is a grain, a contact, a pore chamber, a channel, or a throat. The pore chambers and channels cannot be defined unambiguously, but the ambiguity can be exploited. In problems relating to bulk properties it may be desirable to treat the pore space as composed entirely of pore chambers, whereas in those problems relating to transport through the pore space it may be desirable to treat the pore space as comprised entirely of channels.

Whichever the more convenient partitioning of pore space, one is led back ultimately to representation of the pore space by a graph which is a deformation retract of the pore space for the case of pore chambers only, but need not be for the case of channels only. We have briefly presented the underlying topology and explicit procedures for obtaining deformation retracts both of the grain and the pore spaces. We have shown that these deformation retracts are useful not only for describing the topology of the spaces, but also for detailed quantitative studies of the macroscopic properties of detrital sedimentary rocks. Questions such as those we started with in the introduction have either been answered in part, viz. the one about nuclear magnetic relaxation, or the problems have received a new and illuminating formulation, viz. those about flow and frame moduli.

VI. ACKNOWLEDGEMENT

This work was supported in part by the National Science Foundation, Grant NSF DMR 802069. A part of the work was done while one of us (MHC) visited the Schlumberger-Doll Research Center.

REFERENCES

1. M. H. Cohen and K. Mendelson, submitted to J. Appl. Phys.
2. P. N. Sen, C. Scala, and M.H. Cohen, Geophysics $\underline{46}$, 781 (1981).
3. K. Mendelson and M. H. Cohen, Geophysics (in press).
4. P. N. Sen, this volume.
5. cf. D. Johnson or J. Berryman, this volume.
6. W. F. Brace, J. Geophys. Res. $\underline{82}$, 3343 (1977).
7. C. Lin and M. H. Cohen, submitted to J. Appl. Phys.
8. L. K. Barrett and C. S. Yust, Metallography $\underline{3}$, 1 (1970).
9. F. N. Rhines, Stereology, Springer-Verlag, New York (1967).
10. G. Matheron, Elements pour une theorie des Milieux Poreux, Masson, Paris (1967).
11. E. R. Davis and A.P.N. Plummer, BPRA Conference on Pattern Recognition, Oxford (1980).
12. J. Freer, J. Mol. Biol. $\underline{82}$, 279 (1974).
13. W. K. Pratt, Digital Image Processing, Wiley, New York (1978).
14. A. Rosenfeld, J. ACM $\underline{17}$, 146 (1970).
15. A. C. Shaw, J. ACM $\underline{17}$, 453 (1970).
16. J. Serra, Leitz Sci. Tech. Inform. Supplement 1, 4, p. 125, Wetzler, Apr. (1974).
17. M. Rink and J. R. Schopper, Pageoph 114 (1976).
18. J. Koplick, Creeping Flow in Two-Dimensional Networks, preprint.

WAVE PROPAGATION IN BUBBLY LIQUIDS

Donald A. Drew
U.S. Army Research Office and
Rensselaer Polytechnic Institute

and

Lap-Yan Cheng
Rensselaer Polytechnic Institute
Troy, New York 12181

Wave propagation in a bubbly liquid is studied from the point of view of determining the validity of two-fluid two-phase flow models and the values of various parameters appearing in them. At low frequencies, the speed of sound depends on the virtual mass coefficient. At higher frequencies, the results indicate the need for a virtual heat capacity.

I. INTRODUCTION

Bubbly flows occur in many practical technological situations, including chemical reactors and nuclear reactor cooling systems. An ability to predict system behavior during flow transients is extremely desirable. The flow geometry, including the motion and vibrations of the interfaces, is complex, but for most practical purposes only certain averaged quantities (such as average heat transfer to or from the bubbles) are of interest. For this reason, averaging methods have been applied to the equations of motion for the constituent continua with the hope that some insight can be gained into the macroscopic, or averaged, motions. The success of this approach has been modest, with few results coming from the application of averaging to a realistic flow situation. Consideration of the analogous process of deriving the equations of fluid dynamics from the Boltzmann equation suggests that success at this step is not essential for obtaining useful fluid-like models for the macroscopic motions.

The approach which seems most fruitful in obtaining a working two-phase flow model is one which combines a postulational, or constitutive, approach for the averaged motions with an empirical, or rheological, approach to verify the postulated forms and to measure the coefficients which arise in the constitutive equations. This procedure is analogous to postulating that a (single-phase) fluid is Newtonian (linearly viscous), then verifying that the shear stress is a linear function of the rotation rate in a couette viscometer, and finally extracting the viscosity from the slope of the stress-rotation curve. This procedure replaces the derivation of closure relations from microscopic considerations with deduction of appropriate relations between the macroscopic variables. Intuition and empiricism are strong components of this approach; nevertheless, there is a logical framework within which certain simplifications occur (Drew & Lahey 1979).

In this approach, the features which result from the random (or more appropriately, heterogeneous) nature of the medium are contained in the constitutive equations. Indeed, the difference between bubbly flow and some other flow, for example, droplet flow, are contained in the equations expressing how the two media interact, and in an empirical statement about range of applicability.

Because two-phase flow involves the description of the motions of two separate materials, it is more complicated than single phase flow. In addition to consideration of the flux terms (stresses and heat flux) which are also present in single phase flows, it is necessary to consider terms expressing the interactions between the two phases. Moreover, a cursory examination of virtual mass effects for bubbly flows (Drew, Cheng and Lahey 1979) shows that non-steady effects in the interaction terms can be bigger than the initial terms in the bubble momentum equation.

Propagation of infinitesimal disturbances (sound) represents one way to verify and quantify some of the nonsteady aspects of the interactions between phases. There are many studies of sound propagation, almost invariably done so that the basic flow has no motion (see, e.g. Van Wijngaarden). In subsequent sections, we shall discuss a model which will alow non-zero basic flows. We will also discuss various aspects of the constitutive assumptions; we shall linearize the equations about the state of uniform bubbly flow and calculate the propagation modes for continuous waves (CW).

II. GENERAL FRAMEWORK

The time-averaged equations of motion for two-phase flow are given by Ishii (1975) as

$$\frac{\partial \alpha_k \rho_k}{\partial t} + \nabla \cdot \alpha_k \rho_k \underline{v}_k = \Gamma_k \tag{1}$$

$$\frac{\partial \alpha_k \rho_k \underline{v}_k}{\partial t} + \nabla \cdot \alpha_k \rho_k \underline{v}_k \underline{v}_k = -\nabla \alpha_k p_k + \nabla \cdot \alpha_k \underline{\underline{T}}_k + \Gamma_k \underline{v}_{ki} + \underline{M}_k^d + \underline{M}_k^p \tag{2}$$

$$\frac{\partial \alpha_k \rho_k (e_k + \tfrac{1}{2}v_k^2)}{\partial t} + \nabla \cdot \alpha_k \rho_k (e_k + \tfrac{1}{2}v_k^2) \underline{v}_k = -\nabla \cdot \alpha_k p_k \underline{v}_k + \nabla \cdot \alpha_k \underline{\underline{T}}_k \cdot \underline{v}_k$$

$$-\nabla \cdot \alpha_k \underline{q}_k + \Gamma_k (e_{ki} + \tfrac{1}{2}v_{ki}^2) + E_k \quad (3)$$

where α_k is the volume fraction of phase k, ρ_k is the density, $\underline{v}_k$ is the velocity, Γ_k is the rate of creation of phase k, p_k is the pressure, $\underline{\underline{T}}_k$ is the extra (viscous plus Reynolds) stress, $\underline{v}_{ki}$ is the interfacial average velocity, $\underline{M}_k^d + \underline{M}_k^p$ interphase force, e_k is the energy density, $\underline{q}_k$ is the heat flux, e_{ki} is the interfacial average energy density, and E_k is the interfacial heat transfer rate.

The jump conditions are

$$\Gamma_1 + \Gamma_2 = 0 \, , \quad (4)$$

$$\Gamma_1 (\underline{v}_{1i} - \underline{v}_{2i}) + \underline{M}_1^d + \underline{M}_2^d + \underline{M}_1^p + \underline{M}_2^p = \underline{M}_m \, , \quad (5)$$

$$\Gamma_1 (e_{1i} - e_{2i} + \tfrac{1}{2}v_{1i}^2 - v_{2i}^2) + E_1 + E_2 = E_m \, , \quad (6)$$

where $\underline{M}_m$ and E_m are the mixture sources of momentum and energy at the interface, respectively, and represent the contributions due to surface tension.

For concreteness, we shall denote the bubble phase by a subscript g, and the liquid phase by a subscript ℓ. Furthermore, we write $\alpha_g = \alpha$, and $\alpha_\ell = 1-\alpha$.

III. CONSTITUTIVE EQUATIONS

These equations (1-6) must be supplemented with constitutive equations expressing Γ_k, $\underline{v}_{ki}$, $\underline{M}_k^d$, $\underline{\underline{T}}_k$, $\underline{q}_k$, E_k, and e_{ki} in terms of α, $\underline{v}_k$, p_k and T_k (Drew and Lahey 1979). First, we ignore the macroscopic viscous terms:

$$\underline{\underline{T}}_k = 0 \, . \quad (7)$$

We further assume that

$$\Gamma_k = 0 \, . \quad (8)$$

We assume that the pressure contribution to the interfacial momentum transfer can be represented as

$$M_k^p = p_{ki} \nabla \alpha_k \quad , \tag{9}$$

where p_{ki} is the interfacial averaged pressure. We assume an interfacial force model of the form

$$\underline{M}_g^d = (\underline{F}_d + \underline{F}_{vm}) \quad ,$$

where

$$\underline{F}_d = \frac{3}{8} \rho_\ell \frac{C_D}{r} |\underline{v}_g - \underline{v}_\ell| (\underline{v}_g - \underline{v}_\ell) \quad , \tag{10}$$

and

$$\underline{F}_{vm} = C_{vm} \rho_\ell \left[\left(\frac{\partial \underline{v}_\ell}{\partial t} + \underline{v}_g \cdot \nabla \underline{v}_\ell \right) - \left(\frac{\partial \underline{v}_g}{\partial t} + \underline{v}_\ell \cdot \nabla \underline{v}_g \right) \right.$$

$$\left. + (1-\lambda)(\underline{v}_\ell - \underline{v}_g) \cdot \nabla (\underline{v}_\ell - \underline{v}_g) \right] \tag{11}$$

Here C_D is the drag coefficient and r is the average radius associated with the bubbles. The drag coefficient C_D is assumed to be a function of α and the local Reynolds number Re. Also, C_{vm} is the virtual volume coefficient, and λ is a parameter. Both C_{vm} and λ depend on α.

We shall further take

$$\underline{M}_m = -\frac{2\sigma}{r} \nabla \alpha \quad , \tag{12}$$

where σ is the coefficient of surface tension. By assuming that

$$p_{gi} - p_{\ell i} = \frac{2\sigma}{r} \quad , \tag{13}$$

equation (5) becomes

$$\underline{M}_g^d = -\underline{M}_\ell^d \quad . \tag{14}$$

The task of relating the interfacial pressures p_{ki} to the average pressures p_k remains. For the bubbles, we shall assume

$$p_g = p_{gi} \quad . \tag{15}$$

It is necessary to account for scattering of pressure waves into the liquid. For this purpose, we shall assume that the scattering effects occur at a scale smaller than the interbubble distance. That is, we assume that pressure fluctuations in the neighborhood of the bubble are spherically symmetric. This leads to the Rayleigh equation

$$P_\ell - P_{\ell i} = -\left\{\rho_\ell \left[r \frac{D_g^2 r}{Dt^2} + \frac{3}{2}\left(\frac{D_g r}{Dt}\right)^2\right] + \frac{\mu}{r}\frac{D_g r}{Dt}\right\}, \quad (16)$$

where $D_g/Dt = \partial/\partial t + \underline{v}_g \cdot \nabla$, and μ is the effective viscosity of the surrounding liquid. It is essential to include the Rayleigh equation in order to capture the effects of bubble resonance.

Let us now discuss the thermodynamic assumptions. For the bubbles, we shall assume equations of state of the form

$$P_g = \rho_g R T_g, \quad (17)$$

where T_g is the gas temperature, and R is the gas constant for air. Also, we write

$$h_g = e_g + P_g/\rho_g. \quad (18)$$

We shall assume

$$h_g = c_g T_g \quad (19)$$

where h_g is the gas enthalpy, and c_g is the specific heat of the gas at constant pressure. For the liquid, we write

$$\rho_\ell = \rho_\ell(P_\ell, h_\ell) \quad (20)$$

with

$$h_\ell = e_\ell - P_\ell/\rho_\ell. \quad (21)$$

For the sound propagation studies, we need only the linearized version of equation (20). This becomes

$$\delta\rho_\ell = \frac{\partial \rho_\ell}{\partial P_\ell} \delta P_\ell + \frac{\partial \rho_\ell}{\partial h_\ell} \delta h_\ell$$

$$= \beta_k \delta P_\ell + \phi_k \delta h_\ell \quad (22)$$

where β_k is related to the compressibility of the liquid, and ϕ_k is related to the thermal expansion. Both quantities are standard properties, and are well known for water.

We shall ignore the macroscopic heat flux:

$$\underline{q}_k = 0. \quad (23)$$

The interfacial heat transfer terms E_k needed a more complicated model than first anticipated. We assume that E_m is negligible, and write

$$E_\ell = -E_g = \alpha H(T_{\ell i} - T_\ell) \qquad (24)$$

where H depends on the thermodynamic properties of the two phases and the interface geometry. For low frequencies, equation (24) worked well with a standard model for the coefficient H. For high frequencies (near bubble resonance), however, a model was derived from microscopic considerations after linearization of the fields involved. Since the analysis relates directly to the kinds of constitutive assumptions needed for an adequate description of two-phase media, it will be summarized here.

The harmonically oscillating linearized (radial) continuity, momentum and energy equation in the gas bubble were solved subject to the temperature and heat flux boundary conditions,

$$\hat{T}_\ell(r_o,t) = \hat{T}_g(r_o,t)$$

$$-\kappa_\ell \frac{\partial \hat{T}_\ell(r_o,t)}{\partial r} = -\kappa_g \frac{\partial \hat{T}_g(r_o,t)}{\partial r}$$

and the kinematic condition,

$$\hat{u}(r_o,t) = \frac{dr}{dt},$$

where r_o is the undisturbed bubble radius, and κ_g and κ_ℓ are the thermal conductivities of the gas and liquid, respectively. Note that $\hat{T}_k$ and $\hat{u}$ represent the exact temperatures and the interface velocity, respectively. The analysis extends the work of Plesset and Hsieh (1960) and of Prosperetti (1977). The resulting heat transfer coefficient H was complex valued, and frequency dependent.

IV. WAVE PROPAGATION

In order to study the propagation of infinitesimal disturbances, we assume that the variables depend only on z and t, and that all velocities have only non-zero components in the z-direction. Thus,

$$\underline{v}_k = u_k(z,t)\underline{e}_z . \qquad (25)$$

The previously derived conservation laws, which describe a two-phase mixture as a two-fluid medium with different pressures in each phase (gas/liquid) can be written as a system of nonlinear first order partial differential equations. After linearization, these conservation laws take the form

$$\underline{\underline{A}}(\underline{\phi})\frac{\partial \delta \underline{\phi}}{\partial t} + \underline{\underline{B}}(\underline{\phi})\frac{\partial \delta \underline{\phi}}{\partial z} = \underline{\underline{C}}(\underline{\phi})\delta\underline{\phi} \qquad (26)$$

where $\underline{\phi}$ is the vector of all the field variables: α, p_ℓ, u_g, u_ℓ, h_ℓ, $\dot{r}$, r. The perturbations in gas pressure and gas enthalpy, δp_g and δh_g, are related to δr by expressions derived in an analysis of the heat transfer between a pulsating bubble and the surrounding liquid.

If we assume a traveling wave perturbation

$$\delta\underline{\phi} = \underline{\xi}\, e^{i(kz-\omega t)} \quad , \qquad (27)$$

then Equation (26) becomes

$$\left\{ \underline{\underline{A}}(\underline{\phi})[-i\omega] + \underline{\underline{B}}(\underline{\phi})[ik] - \underline{\underline{C}}(\underline{\phi}) \right\} \delta\underline{\phi} = 0 \quad , \qquad (28)$$

In order for the $\delta\underline{\phi}$'s to be finite, we must have,

$$\det\left\{ (\omega\underline{\underline{A}} - i\underline{\underline{C}}) - k\underline{\underline{B}} \right\} = 0 \quad . \qquad (29)$$

Equation (24) is the dispersion relation from which, for the given ω's, we can solve for the k's. This gives the dispersion relationship, that is the wave number's (k) dependence on the frequency (ω),

$$k = k(\omega) \qquad (30)$$

One physical problem of interest is the boundary value problem modeling the situation where a transducer is located at the bottom of the flow and sound is propagated upward through the bubbly liquid. If the transducer is driven sinusoidally, the resulting pressure disturbances should reach an equilibrium situation where the only frequency observed is the driving frequency. This continuous wave (CW) flow situation is expressed as the boundary value problem

$$\delta p_\ell(0,t) = a\cos\omega t \quad , \qquad (31)$$

$$\delta p_\ell(L,t) = 0 \quad . \qquad (32)$$

All other perturbations at z=0 are 0, except $\delta\alpha$. Equation (31) expresses the action of the transducer, while equation (32) states that the top (z=L) is open to the atmosphere. The significance of the remaining boundary conditions is that the gas and liquid are being injected at specified velocity, but that the "flow split" (the ratio of gas to liquid mass flow rates) is not specified.

Since the channel is long, we approximate the solutions with

$L \to \infty$. This removes the possibility of resonance with the channel, but allows us to ignore modes which have Im $k < 0$.

The solutions for $\delta\phi$ have the form,

$$\delta\phi = \sum_n C_n \xi_n e^{ik_n z} e^{-i\omega t} + \begin{bmatrix} \text{Complex} \\ \text{Conjugate} \end{bmatrix}, \qquad (33)$$

where k_n is the (complex) wave number of the nth mode and ξ_n the corresponding eigenvector for Equation (28). Because of the condition discussed above $C_n = 0$ for all k_n's having negative imaginary parts. The relative importance of each harmonic (corresponding to each eigenvalue k) in the solution of Equation (33) is based upon the relative magnitude of C_n; the greater the value, the more the importance. Using Equations (31-33) we can solve for the C_n's. The mode which dominates will be the one which is actually measured in standing-wave experiments.

V. RESULTS AND CONCLUSION

The high frequency data which we used for comparison are those due to Silberman (1957). As can be seen in Figures 1 and 2, the dispersion model appears to predict the (wave) phase velocity and attenuation quite well for frequencies both above and below bubble resonance. In Figure 2, virtual mass is observed to have a noticeable effect on the phase velocity at frequencies around the resonance. The resonant spike seen in the attenuation plot is due to insufficient damping. Although there isn't enough data to be conclusive, it can be noted in Figure 1 that the predicted attenuation at the resonance point appears to be closer to the data when virtual mass is absent. However, it is possible that there are other dissipation mechanisms at resonance which we have not modelled (e.g., finite amplitude effects, relaxation effects, etc.). Thus, it is premature to draw strong conclusions from these plots. The low frequency data are taken from Hall's (1971) thesis. As can be noted in Figure 3, the effect of virtual mass is more significant in Hall's data than in Silberman's data, apparently because the void fractions in Hall's experiments are about 200 times higher than in Silberman's data. Note in Figure 3 that if virtual mass is not included the agreement is very poor. As can be seen in Figure 3-5, $C_{VM} = 0.5$ appears to agree with the data.

Unfortunately, there is too much scatter in the data to be able to give a good functional form for $C_{VM}(\alpha)$. Moreover, other experimentation showed no sensitivity at all to different values of λ.

Fig. 2: Attenuation vs. frequency, same conditions as Fig. 1.

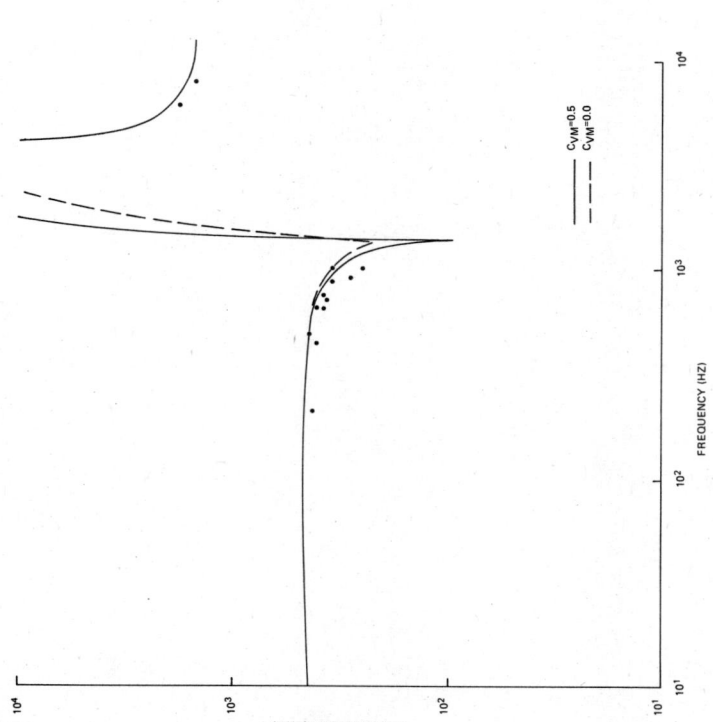

Fig. 1: Phase velocity vs. frequency, $r_o = 2.5 \times 10^{-3}$ m., $\alpha_o = 5.84 \times 10^{-4}$

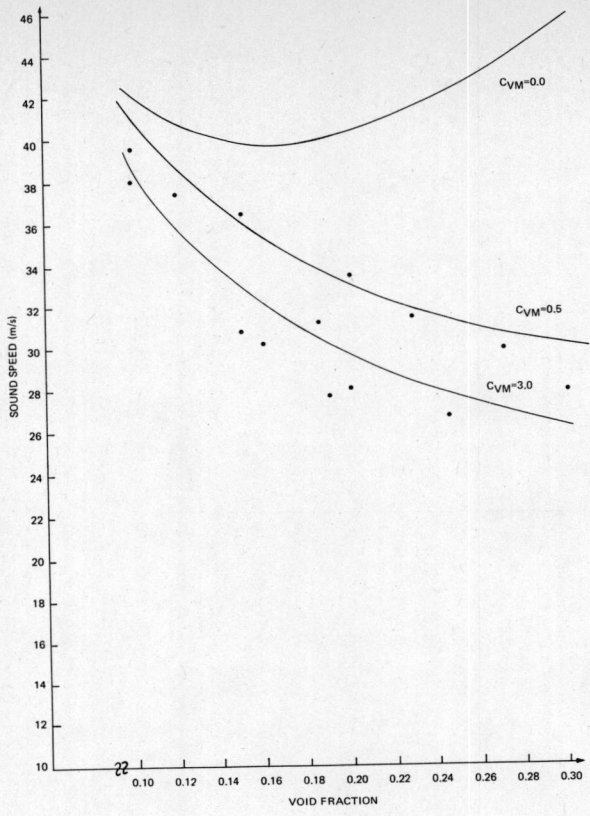

Fig. 3: Sound speed vs. void fraction. Frequency 30 Hz.

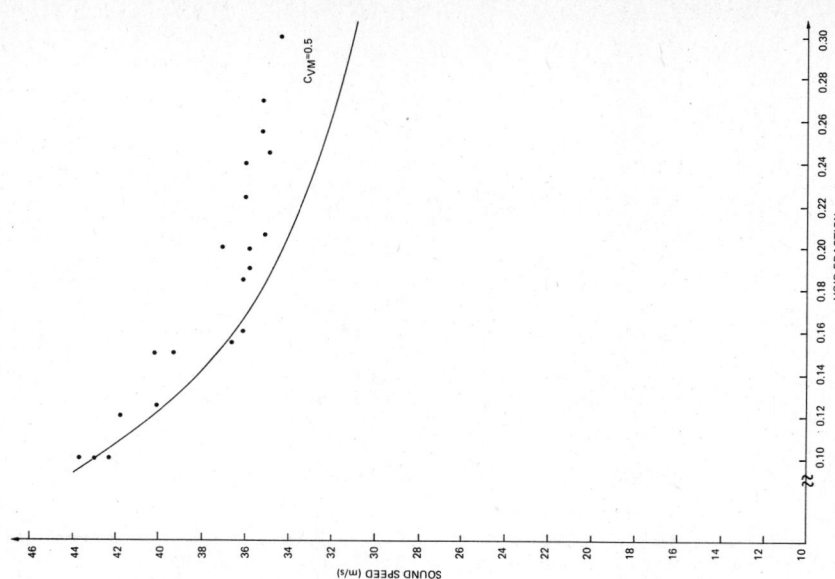

Fig. 5: Sound speed vs. void fraction. Frequency 100 Hz.

Fig. 4: Sound speed vs. void fraction. Frequency 50 Hz.

The implications of the lack of success with a steady-state model for the interfacial heat transfer parameter H have not been fully evaluated. One tentative conclusion is that a "virtual heat capacity" is needed in the interfacial heat transfer model.

REFERENCES

Drew, D.A., L.Y. Cheng, and R.T. Lahey, Jr. (1979). The analysis of virtual mass effects in two-phase flow, Int. J. Multiphase Flow 5, 233-242.

Drew, D.A. and R.T. Lahey, Jr. (1979). Application of general constitutive principles to the derivation of multidimensional two-phase flow equations, Int. J. Multiphase Flow 5, 243-264.

Hall, P. (1971). The propagation of pressure waves and critical flow in two-phase mixtures. Ph.D. Thesis, Heriot-Watt University, Edinburgh, G.B.

Ishii, M. (1975). Thermo-Fluid Dynamic Theory of Multiphase Flow, Eyrolles, France.

Plesset, M.S. and D.Y. Hsieh (1960). Theory of gas bubble dynamics in oscillating pressure fields, Phys. of Fluids 3, 6, 882-892.

Prosperetti, A. (1977). Thermal effects and damping mechanisms in the forced radial oscillations of gas bubbles in liquids, J. Acoustic Soc. Am. 61, 1, 17-27.

Silberman, E. (1957). Sound velocity and attenuation in bubbly mixtures measured in standing wave tubes, J. Acoustic Soc. Am. 29, 925-933.

Wijngaarden, L. van (1972). One dimensional flow of liquids containing small gas bubbles, Ann. Rev. Fluid Mech., vol. 4.

ELASTODYNAMICS OF POROUS MEDIA

David Linton Johnson
Schlumberger-Doll Research
P. O. Box 307
Ridgefield, CT 06877

ABSTRACT

The viscous and elastic properties of a two component medium are derived from a general 2 component Lagrangian formulation. For the case when one of the components is a fluid, there are 2 compressional waves, one of which (the slow wave) is diffusive at low frequencies and propagatory at high. Comparison is made against experiments on various porous fluid saturated systems. It is seen that the diffusive mode in polymer gels, 4th sound in HeII, the diffusion of a fluid pressure pulse through a porous medium, and the recent observation of a slow compressional propagatory wave in water saturated fused glass beads are special cases of this additional mode prediced by the theory.

It is generally recognized that an important aspect of acoustic attenuation and dispersion in porous fluid saturated media is due to the relative motion that can occur between the solid and the fluid parts. Therefore, it is important to develop a theory in which the displacements of the two components are followed separately and on an equal footing, at least in some macroscopic, average sense. A theory of this type was developed by Biot[1-4] and we will explore its consequences (below) but first let us consider the most general continuous Lagrangian formulation of wave mechanics in a two component system where it is envisioned that there exist volume elements dV small compared to the relevant wave length but large compared to the individual grains of the two components; the Biot theory will seem to be a special case of this formulation. Let $\vec{U}^a(\vec{r},t)$ be the average displacement of component a within dV at position $\vec{r}$ and time t. Define the symmetric and anti-symmetric strain tensors for each component:

$$\epsilon_{KL}^a = \frac{1}{2}\left[\frac{\partial U_K^a}{\partial x_L} + \frac{\partial U_L^a}{\partial x_K}\right] \quad a = 1,2$$

$$\omega_{KL}^a = \frac{1}{2}\left[\frac{\partial U_K^a}{\partial x_L} - \frac{\partial U_L^a}{\partial x_K}\right]$$

If one wishes a wave-like equation of motion i.e. with no derivatives higher than 2nd order, then the Lagrangian contains terms up to first order in derivatives only. We insist that it costs no energy for a uniform rotation or translation of both components. The most general Lagrangian density for a homogeneous, isotropic system involving two components is, therefore[5]

$$\mathcal{L} = \frac{1}{2} \{\rho_{11}|\dot{\vec{U}}_1|^2 + 2\rho_{12}\dot{\vec{U}}_1\cdot\dot{\vec{U}}_2 + \rho_{22}|\dot{\vec{U}}_2|^2$$

$$-[\gamma_1\epsilon^1_{KK}\epsilon^1_{LL} + \gamma_2\epsilon^1_{KK}\epsilon^2_{LL} + \gamma_3\epsilon^2_{KK}\epsilon^2_{LL}$$

$$+\gamma_4\epsilon^1_{KL}\epsilon^1_{KL} + \gamma_5\epsilon^1_{KL}\epsilon^2_{KL} + \gamma_6\epsilon^2_{KL}\epsilon^2_{KL}$$

$$+\gamma_7(\omega^1_{KL} - \omega^2_{KL})(\omega^1_{KL} - \omega^2_{KL}) + \gamma_8(\vec{U}_1 - \vec{U}_2)^2]\}$$

$$= \frac{1}{2} \{\rho_{11}|\dot{\vec{U}}_1|^2 + 2\rho_{12}\dot{\vec{U}}_1\cdot\dot{\vec{U}}_2 + \rho_{22}\dot{\vec{U}}_2\cdot\dot{\vec{U}}_2$$

$$-[\alpha_1(\nabla\cdot\vec{U}_1)^2 + 2\alpha_2(\nabla\cdot\vec{U}_1)(\nabla\cdot\vec{U}_2) + \alpha_3(\nabla\cdot\vec{U}_2)^2$$

$$+\alpha_4|\nabla\times\vec{U}_1|^2 + 2\alpha_5(\nabla\times\vec{U}_1)\cdot(\nabla\times\vec{U}_2) + \alpha_6|\nabla\times\vec{U}_2|^2$$

$$+\nabla\cdot\vec{A}_T + \alpha_8(\vec{U}^1 - \vec{U}^2)^2]\} \; .$$

where the set $\{\rho_{ij}\}$, $\{\gamma_i\}$, $\{\alpha_i\}$ are parameters of the system. (Each α_i is linearly related to the γ_i's.) In general, one does not know how to calculate these parameters. The relevance of the 2nd form is that the term $\nabla\cdot\vec{A}^T$ drops out identically of the equations of motion, which are:

$$\rho_{11}\frac{\partial^2\vec{U}^1}{\partial t^2} + \rho_{12}\frac{\partial^2\vec{U}^2}{\partial t^2} = \alpha_1\nabla(\nabla\cdot\vec{U}^1) + \alpha_2\nabla(\nabla\cdot\vec{U}^2)$$

$$-\alpha_4\nabla\times(\nabla\times\vec{U}^1) - \alpha_5\nabla\times(\nabla\times\vec{U}^2)$$

$$-\alpha_8(\vec{U}^1 - \vec{U}^2)$$

$$\rho_{22}\frac{\partial^2\vec{U}^2}{\partial t^2} + \rho_{12}\frac{\partial^2\vec{U}^1}{\partial t^2} = \alpha_3\nabla(\nabla\cdot\vec{U}^2) + \alpha_2\nabla(\nabla\cdot\vec{U}^1)$$

$$-\alpha_6\nabla\times(\nabla\times\vec{U}^2) - \alpha_5\nabla\times(\nabla\times\vec{U}^1)$$

$$-\alpha_8(\vec{U}^2 - \vec{U}^1)$$

The plane wave normal modes can almost be determined by inspection. There are two longitudinal modes and two transverse modes (of each polarization) and the dispersion relations are

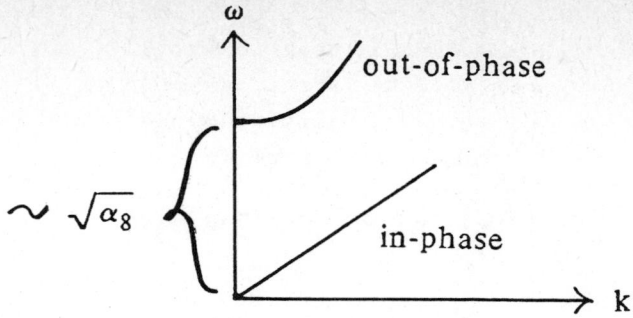

Fig. 1
Acoustic dispersion relation for either transverse
or longitudinal modes of a 2 component medium.

The upper branch is the formal equivalent of an optic phonon branch whereas the lower is the acoustic branch. The frequency "cutoff" of the upper branch is a direct manifestation of the restoring force due to uniform displacement of the two components with respect to each other.

If one specializes to the case where one of the components ($\vec{U}^2$ say) is a fluid, then α_8 is identically zero because if the 2 components are uniformly displaced relative to each other, they don't try to "snap back". Moreover, we shall postulate that the fluid does not experience any shear restoring force, nor does it contribute to one on the solid ($\alpha_5 = \alpha_6 = 0$). In that case *both* longitudinal modes are non-dispersive and there is only one shear mode (the other shear mode is an identically zero frequency mode for all wave vectors).

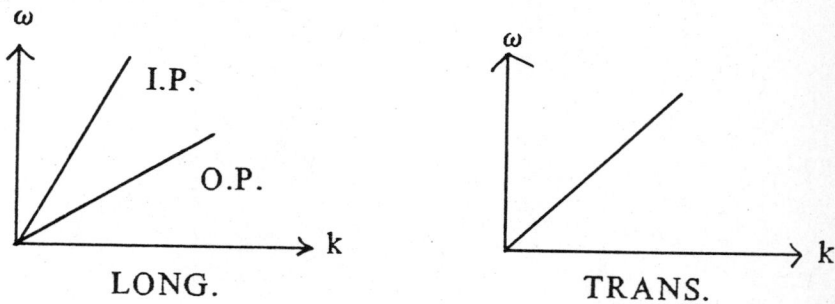

Fig. 2
Longitudinal (left) and transverse (right) dispersion relation
of a fluid-solid medium. Fluid is assumed to be non-viscous.

Note that the "slow" compressional, (longitudinal) mode always corresponds to $\vec{U}^1$ and $\vec{U}^2$ moving out of phase (O.P.) with respect to each other. Up to this point an important piece of physics has been neglected, namely the viscous attenuation created by the relative fluid-solid motion. This can be introduced into the Lagrangian formation via a dissipation function but we shall, for the purposes of the rest of this paper, use the formulation due to Biot. We shall also use his notation.

In a series of papers,[1-4] Biot proposed a simple phenomenological theory of acoustic propagation in porous, fluid filled, macroscopically homogeneous and isotropic media. It is assumed that there exist volumes large compared to pore/grain sizes but small compared to a wavelength and that each volume element is describable by the average displacement of the fluid $\vec{U}(\vec{r},t)$ and of the solid parts $u(\vec{r},t)$. The equations of motion, including viscous damping, are

$$\rho_{11}\frac{\partial^2 \vec{u}}{\partial t^2} + \rho_{12}\frac{\partial^2 \vec{U}}{\partial t^2} = P\nabla(\nabla\cdot\vec{u}) + Q\nabla(\nabla\cdot\vec{U}) - N\nabla\times\nabla\times\vec{u}$$

$$+ bF(\omega)\left[\frac{\partial \vec{U}}{\partial t} - \frac{\partial \vec{u}}{\partial t}\right] \tag{1a}$$

$$\rho_{22}\frac{\partial^2 \vec{U}}{\partial t^2} + \rho_{12}\frac{\partial^2 \vec{u}}{\partial t^2} = R\nabla(\nabla\cdot\vec{U}) + Q\nabla(\nabla\cdot\vec{u})$$

$$- bF(\omega)\left[\frac{\partial \vec{U}}{\partial t} - \frac{\partial \vec{u}}{\partial t}\right] \tag{1b}$$

Here we have used the notation of ref. 1 as it is simpler than (although equivalent to) that of the later articles. P, Q, R are generalized elastic coefficients which can be related[1,4] to the bulk modulus of fluid K_f, the bulk modulus of solid K_s, the bulk modulus K_b of the skeletal frame ("jacketed and drained") and to N which is the shear modulus of both the skeletal frame and of the composite:

$$P = \frac{(1-\phi)\left[1-\phi-\frac{K_b}{K_s}\right]K_s + \phi\frac{K_s}{K_f}K_b}{1-\phi-\frac{K_b}{K_s}+\phi\frac{K_s}{K_f}} + \frac{4}{3}N \quad, \tag{2a}$$

$$Q = \frac{\left[1-\phi-\frac{K_b}{K_s}\right]\phi K_s}{1-\phi-\frac{K_b}{K_s}+\phi\frac{K_s}{K_f}} \quad, \tag{2b}$$

$$R = \frac{\phi^2 K_s}{1-\phi-\frac{K_b}{K_s}+\phi\frac{K_s}{K_f}} \quad. \tag{2c}$$

ϕ is the porosity (fluid volume fraction). In the so-called "jacketed and drained" gedanken test[4] the sample is stressed but the fluid is allowed to escape as needed in order to remain at ambient pressure. Therefore, K_b and N are the elastic constants of the bare skeletal frame and, in the absence of an electrochemical interfacial effect between fluid and solid, they are independent of what fluid is in the pores, including vacuum; we will make this assumption throughout the paper.[6] Equations (2) are equivalent to those given in e.g., Stoll,[7] Geertsma and Smit.[8] The density terms ρ_{ij}, are related to the density of solid, ρ_s, and fluid, ρ_f, by

$$\rho_{11} + \rho_{12} = (1-\phi)\rho_s , \tag{3a}$$

$$\rho_{22} + \rho_{12} = \phi \rho_f . \tag{3b}$$

The term ρ_{12} describes the inertial (as opposed to viscous) drag that the fluid exerts on the solid as the latter is accelerated relative to the former and vice-versa. The equation of motion of the solid part, for example, equation (1a), may be rewritten using (3a):

$$(1-\phi)\rho_s \frac{\partial^2 \vec{u}}{\partial t^2} = -\rho_{12}\left[\frac{\partial^2 \vec{U}}{\partial t^2} - \frac{\partial^2 \vec{u}}{\partial t^2}\right]$$

$$+ bF(\omega)\left[\frac{\partial \vec{U}}{\partial t} - \frac{\partial \vec{u}}{\partial t}\right] + \text{(spatial derivative terms)} \tag{1a'}$$

That is, even for a non-viscous pore fluid ($bF(\omega)\equiv 0$), there is a reactive force per unit volume on the solid [whose mass is $(1-\phi)\rho_s$] whenever one component is accelerated relative to the other. The proportionality constant, ρ_{12}, represents the induced mass tensor[9] per unit volume, assumed to be diagonal in the coordinate indices for a homogeneous isotropic system; it is always negative and is always proportional to the fluid density:

$$\rho_{12} = -(\alpha-1)\phi\rho_f \tag{3c}$$

where $\alpha > 1$ is a purely geometrical quantity independent of solid or fluid densities. Berryman[10] has considered the case of isolated spherical solid particles in the fluid to derive $\alpha = (1/2)[\phi^{-1}+1]$, for example. The remaining parameters govern attenuation; $b = \eta\phi^2/k$ where η is the fluid viscosity, k is the permeability, and $F(\omega)$ allows for the fact that the effective damping changes when the viscous skin depth ($\sqrt{2\eta/\rho_f\omega}$) becomes smaller than the pore size as the frequency ω increases.[11] It is worth noting that ρ_{12} and $bF(\omega)$ *always* appear together in the combination $\rho_{12} - \frac{ibF(\omega)}{\omega}$. i.e., equation (1a') can be rewritten

$$(1-\phi)\rho_s \frac{\partial^2 \vec{u}}{\partial t^2} = -\tilde{\rho}_{12}(\omega)\left[\frac{\partial^2 \vec{U}}{\partial t} - \frac{\partial^2 \vec{u}}{\partial t^2}\right] + \text{(spatial derivative terms)} \tag{1a''}$$

where

$$\tilde{\rho}_{12}(\omega) = \rho_{12} - \frac{ibF(\omega)}{\omega}$$

and similarly for equation (1b).

Although the linear term in the Taylors series expansion of $F(\omega)$ mimics the effects of ρ_{12} (i.e., both terms describe an ω^2 dependence of the force on the relative displacement), they are of different physical origin and it is not valid to put $\alpha=1$ (i.e., $\rho_{12}=0$) as has been done.[12] Equivalently, one could define $\tilde{\alpha}(\omega)$ by analogy with equation (3c), viz: $\tilde{\rho}_{12}(\omega) = -[\tilde{\alpha}(\omega)-1]\phi\rho_f$. Therefore,

$$\tilde{\alpha}(\omega) = \alpha + \frac{ibF(\omega)}{\omega\phi\rho_f} \qquad (3c')$$

The crossover between the high-frequency and low-frequency behavior of $F(\omega)$ occurs when the viscous skin depth is approximately equal to the pore size, a:

$$\omega_c = \frac{2\eta}{\rho_f a^2} \ .$$

If $\omega \gg \omega_c$, the attenuation mechanism has little effect on the velocities of the normal modes which are derived from Eqs. (1); one has[1] $\lim_{\omega\to\infty} F(\omega)\alpha\omega^{1/2}$ and therefore $\lim_{\omega\to\infty} \tilde{\rho}_{12}(\omega) = \rho_{12}$ (a constant *not* equal to zero), and $\lim_{\omega\to\infty} \tilde{\alpha}(\omega) = \alpha$ (a constant *not* equal to 1). (As was shown previously,[13] the constant α is the crucial parameter for the slow wave in systems having a very stiff frame; see Equations (8a,b) below also.)

In this high frequency limit, the non-dispersive velocities of the normal modes are

$$V^2(\text{SHEAR}) = \frac{N}{(1-\phi)\rho_s + \phi\rho_f - \frac{1}{\alpha}\phi\rho_f} \ , \qquad (4a)$$

$$V^2(\text{FAST, SLOW}) = \frac{\Delta \pm \sqrt{\Delta^2 - 4(\rho_{11}\rho_{22} - \rho_{12}^2)(PR - Q^2)}}{2(\rho_{11}\rho_{22} - \rho_{12}^2)} \ , \qquad (4b)$$

where

$$\Delta = P\rho_{22} + R\rho_{11} - 2\rho_{12}Q \ .$$

That is, there is one shear wave, a fast compressional wave corresponding to solid and fluid moving in phase, and a slow compressional wave corresponding to solid and fluid moving out of phase. Note that these equations depend on α either explicitly or through Eq. (3c) but not on $bF(\omega)$. Indeed, using various literature approximations to $F(\omega)$ and measured values of $k(\sim 10^{-8} \text{cm}^2)$, we have verified[14] that Eqs. (4a,b) are accurate to better than 1%. (In the ultrasonics experiments considered in this article,[15,16] the viscous skin depth in water is less than 1 μm at 500 kHz whereas the pore sizes are larger than 20 μm. Thus, these experiments were made in the high frequency limit of the theory.)

First, though, we consider the opposite limit, $\omega \ll \omega_c$; the fast and shear waves are still propagatory but the slow wave is described by a diffusion equation,

$$C_D \nabla^2 \vec{\xi} = \frac{\partial \vec{\xi}}{\partial t} \ , \qquad (5a)$$

where $\vec{\xi}$ is a normal mode coordinate and C_D is given by

$$C_D \equiv \frac{kK_f}{\eta\phi}\left[1+\frac{K_f}{\phi\left[K_b+\frac{4}{3}N\right]}\right.$$

$$\left.\cdot\left\{1+\frac{1}{K_s}\left[\frac{4}{3}N\left(1-\frac{K_b}{K_s}\right)-K_b-\phi\left[K_b+\frac{4}{3}N\right]\right]\right\}^{-1}\right] \qquad (5b)$$

(Equations 5 are also derivable from equation 4b by the previously mentioned substitution $\alpha \to \tilde{\alpha}(\omega)$ and then taking the limit $\omega \to 0$.) In ref. 17 it was shown that Eqs.(5a,b) provide the most general description of quasi-static flow in permeable media and that they subsume other theories as special cases (well-to-well pressure pulse testing in the petroleum industry, diffusion in dilute polymer gels). Two cases of special interest (in the low frequency limit) are:

1) Very stiff frame, $K_b, N \gg K_f$. In this case equation 5b simplifies greatly:

$$\lim_{K_b,N \gg K_f} C_D \to \frac{kK_f}{\eta\phi} \qquad (5c)$$

Chandler[17] has measured the diffusivity of porous water saturated media (fused glass bead samples and a Berea sandstone). He found that the permeability deduced from the approximate equation (5c) differs from the statically measured permeability by factors of 2-3 whereas if one accounts for the finite compressibility of the frame (i.e. equation 5b) agreement is much improved.

Fig. 3
Permeability deduced from diffusivity vs. static permeability.

In Fig. 3, G = N, the open symbols refer to permeability deduced from equation (5c) and the closed symbols refer to permeability deduced from the full theory, equation (5b). (The abscissa is the statically measured permeability whereas the ordinate is that deduced from either equation 5b or 5c.) Although one is nearly in the stiff frame limit here, it is not valid to neglect the finite compressibility of the solid matrix.

2) A cross-linked polymer gel in an aqueous solution corresponds to a fluid-solid system in the *weak* frame limit, $K_b, N, \ll K_f, K_s$. Since the pores are considerably less than a micron in size, one is essentially always in the low frequency limit where the slow wave is diffusive. Indeed, light scattering experiments clearly indicate a diffusive mode in addition to the usual propagatory mode.[18] Upon gelation, the individual molecules cross link, thus taking the system from the "sol" phase to the "gel" phase, and so the skeletal frame moduli grow from zero in the former to some small value in the latter. In this weak frame limit the diffusivity Equation 5b simplifies greatly[17,19]:

$$\lim_{K_b, N \ll K_f, K_s} C_D - \frac{k}{\eta}(K_b + \frac{4}{3}N) \qquad (5d)$$

This is to be compared with an expression derived independently by Tanaka et al[18] and DeGennes[20] which is, in the present notation:

$$C_D - \frac{k}{\eta \phi^2}(K_b + \frac{4}{3}N) \qquad (5e)$$

Since (5e) was derived under the assumption of a dilute concentration of molecules ($\phi \approx 1$), it is essentially the same as (5d). We may confidently conclude that the slow wave has been observed in gels. The speed and attenuation of the fast wave (ordinary sound) also change upon gelation. It is straightforward to show from Equation (4b) that the speed in the gel phase is[19]

$$V(FAST) - V_o[1 + \zeta_1 K_b + \zeta_2 N] \qquad (6a)$$

where V_o is the speed in the sol phase and $\{\zeta_i\}$ are complicated functions of ϕ, ρ_f, ρ_s, K_f, K_s. This is to be compared with an expression derived by Bacri, et al.[21] which involved several simplifying assumptions:

$$V(FAST) - V_o[1 + (K_b + \frac{4}{3}N)/(2K_f)]. \qquad (6b)$$

Similarly, the attenuation of the ordinary sound (fast wave) in the weak frame limit is

$$\alpha - \frac{1}{2} \frac{(\rho_T - \rho_f)^2}{V_o \rho_T} \frac{k}{\eta} \omega^2 \qquad (7a)$$

where $\rho_T - (1-\phi)\rho_s + \phi \rho_f$ is the total density of the gel. This is to be compared with an expression derived by Bacri, et al[21]

$$\alpha - \frac{1}{2} \frac{(1-\phi)^2 \rho_s^2}{V_o \rho_T} \frac{k}{\eta} \omega^2 \qquad (7b)$$

which is the same as (7a) *if* one neglects the volume occupied by the polymer molecules, $\rho_T = \rho_f + (1-\phi)\rho_s$. Note that the general Biot theory, equations (1a,b) simultaneously provide a description of the 2 longitudinal modes, which these other theories do not.

Note also that the low frequency limit of the diffusive slow wave has been observed in both the stiff frame (fused glass beads, rocks) and weak frame (polymer gels) limit. It is obviously a fairly common phenomenon.

What happens when the frequency is increased to, say, the point where the viscous skin depth is much less than the pore size? According to the theory $\tilde{\alpha}(\omega) \approx \alpha$, a real constant (Equation 3c') and all 3 modes fast, slow, and shear are propagatory with speeds given by Equations (4a,b). The slow wave had never been observed as a propagatory mode until T.J. Plona[15] of Schlumberger observed it in water saturated fused glass beads and in other porous solids (typically water filter materials). Some of his data are reproduced below[16]

Table I

	Sintered Beads #1	Sintered Beads #2	3M	Mott	Coors
pore size (μm)	–	–	55	20	55
porosity	28.3	18.5	34.5	48.0	41.5
permeability (μm^2)	9.1	1.5	8.8	5.9	8.9
velocity:fast compressional (km/sec)	4.05	4.84	2.76	2.74	3.95
velocity:shear (km/sec)	2.37	2.93	1.41	1.54	2.16
velocity:slow compressional (km/sec)	1.04	0.82	0.91	0.92	0.96
sample thickness (cm)	1.91	1.75	.93	.64	.60
sample diameter (cm)	10.4	10.3	10.8	12.7	10.2

Note that the slow wave speed is always less than that of water (which is 1.5 km/sec).

The theoretical velocities of the modes in the high frequency limit depend on three microgeometric parameters α, K_b, N (in adddition to bulk properties ρ_s, K_s etc.). It is particularly informative to consider the speeds as a continuous function of the frame moduli K_b, N keeping α (and all the other parameters) fixed. For example, an unconsolidated (loose) sample of glass beads corresponds to vanishing frame moduli ($K_b = N = 0$) whereas a light sintering of the beads obviously "stiffen up" the skeletal frame without changing the porosity ϕ or the inertial drag parameter α measurably. We compare the calculated speeds as a function of the longitudinal modulus of the skeletal frame $K_b + \frac{4}{3}N$ in Fig. 4, keeping K_b/N a constant.[22]

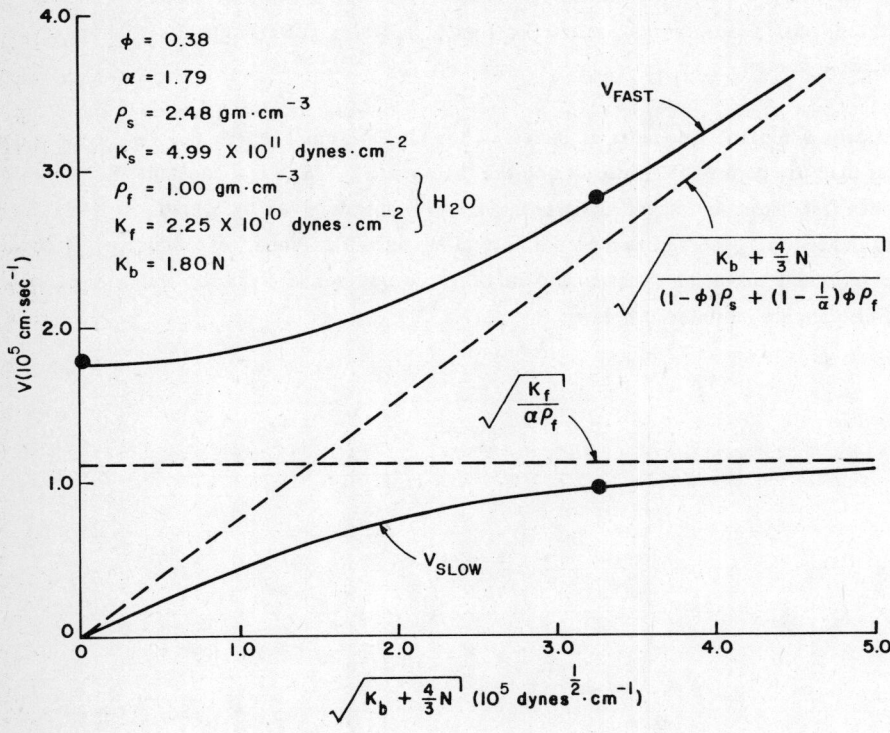

Fig. 4
Fast and slow speeds as a function of frame moduli.

We have also plotted the experimental points corresponding to a lightly fused frame ($K_b, N \neq 0$) and to a non-fused, loose collection of water saturated glass beads ($K_b = N = 0$). We see that in this latter limit both theoretically and experimentally there is only a fast compressional wave whose speeds are in good agreement with each other. We hope eventually to fill in a substantial portion of these curves by controlling the frame moduli; one might, for example, apply an overburden pressure to compress the beads together and thus drive the slow wave speed up from zero.

Notice from Fig. 4 that the complicated expressions for the compressional velocities in the high frequency limit, Equation 4a, simplify greatly if the skeletal frame is very stiff; $K_b, N >> K_f$. In this limit the speeds are[13]:

$$V(\text{FAST}) = \sqrt{\frac{K_b + (4/3)N}{(1-\phi)\rho_s + (1-\alpha^{-1})\phi\rho_f}} \qquad (8a)$$

$$V(\text{SLOW}) = \frac{V_f}{\sqrt{\alpha}} \qquad (8b)$$

The fast wave corresponds now to oscillation of the skeletal frame in which some, but not all, of the fluid is dragged along. The slow wave, on the other hand, corresponds to oscillation of the fluid only; the speed is renormalized down from V_f (speed of sound in fluid) because of the tortuous nature of the pore space. We see from Fig. 4 that the light fusing used to make the consolidated frame already puts us in the stiff frame limit.

Suppose one puts superfluid ^{4}He in the pore space. Since the viscosity of the superfluid fraction is identically zero, we are always in the "high frequency" limit of the theory. Let us compare the prediction of the full theory, Equations 4a,b, with the approximate formulae, Equations (8a,b). The results, for three of Plona's samples, are presented in Table II.[13]

Table II
Theoretical wave speeds for the samples of Table I under the condition of saturation with superfluid ^{4}He.

		Sample 1	Sample 2	Sample 3
	ϕ	0.283	0.258	0.185
	$\alpha = n^2$	1.81	1.94	3.00
Fast-compressional	full theory	4.06	4.18	4.89
speed(km s^{-1})	$\left(\dfrac{K_b + \tfrac{4}{3}N}{(1-\phi)\rho_s + (1-1/\alpha)\phi\rho_f}\right)^{1/2}$	4.06	4.18	4.89
Shear speed (km s^{-1})	full theory	2.44	2.57	3.00
Slow wave/	full theory	173.514	170.797	137.357
fourth-sound speed (m s^{-1})	$\left(\dfrac{K_f}{\alpha\rho_f}\right)^{1/2} = \dfrac{C_4^0(T=0)}{n}$	173.554	170.839	137.381

Since the speeds of the slow wave are very accurately given by $V_f/\sqrt{\alpha}$, we conclude that in this limit the slow wave is identical to the phenomenon known as 4th sound.[23,24] In 4th sound, the normal fluid component (if any) is assumed to be locked relative to the porous superleak by its viscosity and only the superfluid fraction can oscillate; from the 2 fluid equations of motion one can calculate the theoretical speed of 4th sound as a function of temperature, $C_4^0(T)$. The tortuous pore geometry renormalizes the speed so that the experimentally observed speed, $C_4^E(T)$, is considerably less. The ratio of the two is the index of refraction, n, of the 4th sound:

$$C_4^E(T) = \frac{C_4^0(T)}{n} \qquad (9)$$

This equation is identical to Equation (8b) which we have derived from the Biot theory, with $n^2 = \alpha$. Therefore 1) 4th sound, which has been observed since 1965,[24] is an ideal example of the propagatory Biot slow wave. 2) Any porous and permeable solid will exhibit a slow wave if it is saturated with superfluid ⁴He. 3) Observations of the slow wave under condition of water saturation have enabled us to make predictions of 4th sound velocities. Conversely, 4th sound speeds very directly give us the key parameter (α) for the high frequency speed of the slow wave when other fluids are in the pore space.

It is clear that the parameter α, or more generally $\tilde{\alpha}(\omega)$, Equation (3c'), is the key parameter for the slow wave, at least in the theory presented so far. It turns out that α is also simply related to the electrical conductivity of the pore space because both are determined by the solution of Poisson's equation with the same boundary condition.[25] If the (non conducting) matrix is saturated with a conducting fluid of conductivity σ_f, the sample conductivity, σ, is simply $\sigma = \frac{1}{F}\sigma_f$, where the quantity F is a geometrical factor independent of pore fluid condutivity and is related to the acoustic parameter α by $\alpha = F\phi$. A theory,[26] described by Sen in this conference, predicts $F = \phi^{-m}$ where m depends on the aspect ratio of the grains. For spherical grains (m = 3/2) and this result ($F = \phi^{-3/2}$) is perfect agreement with conductivity measurements on fused glass bead samples over a wide range of porosity. Preliminary measurements on the samples used in the acoustics experiments seem to confirm this and will be reported elsewhere.[27] Meanwhile, we have applied[28] the self-similar model to the 4th sound index of refraction data in Fig. 5.

Fig. 5
4th sound indices vs. porosity: theory and experiment

For purposes of the figure, $P = \phi$ (porosity), $n^2 = \alpha$, and $F = \sigma_f/\sigma$. We see that $m = 5/3$ $(n^2 = P^{-2/3})$ corresponding to a random array of needles, gives a good account of the data. This is reasonable considering that the packed powder superleaks used are so fine grained that they stick together to form bridges through the very open, high porosity pore space.

To summarize, we have seen that the slow compressional wave predicted by the theory is a much more commonly observed phenomenon than previously thought. Table III summarizes the systems in which the slow wave has been observed in one form or another.

Table III
Systems in which the slow wave has been observed.

	Weak Frame	Stiff Frame
Low Freq. (diffusive) (dominance of η_f)	Polymer gels	Rocks, fused glass beads. (e.g. Pressure Pulse Testing)
High Freq. (propagatory) (dominance of ρ_f)	Unconsolidated beads under confining pressure (?)	4th sound in HeII Fused glass beads & other artificial media

As yet, a propagatory slow wave in an unconsolidate saturated system (under confining pressure) has <u>not</u> been observed but we expect that it will be. It should be emphasized that it makes no sense whatever to analyze data on the fast compressional and shear waves, in terms of the Biot theory, unless quantitative comparison with the slow wave is also made. Unfortunately, this has not always been the case. Since the theory always predicts a slow wave it remains to explain why it is not always observed.

References

1. M. A. Biot, J. Acoust. Soc. Am. *28*, 168 (1956); *28*, 179 (1956).
2. M. A. Biot, J. Appl. Phys. *33*, 1482 (1962).
3. M. A. Biot, J. Acoust. Soc. Am. *34*, 1254 (1962).
4. M. A. Biot and D. G. Willis, J. Appl. Mech. *24*, 594 (1957).
5. D. L. Johnson (unpublished).
6. This assumption may very well break down when the pore fluid is water and the matrix is a rock [M.R.J. Wyllie, G.H.F. Gardner, and A. R. Gregory, Geophysics *27*, 569 (1962)].
7. R. D. Stoll in *Physics of Sound in Marine Sediments*, edited by L. Hampton (Plenum, NY, 1974).
8. J. Geertsma and D. C. Smit, Geophysics *26*, 169 (1961).
9. L. D. Landau and E. M. Lifshitz, *Fluid Mechanics*, (Pergamon, NY, 1959) p. 31ff.
10. J. G. Berryman, Appl. Phys. Lett. *37*, 382 (1980).
11. Ref. 9, p. 88 ff.
12. C. H. Yew and P. N. Jogi, Exp. Mech. *18*, 167 (1978); H. D. McNiven and Y. Mengi, J. Acoust. Soc. Am. *61*, 972 (1977).
13. D. L. Johnson, Appl. Phys. Lett. *37*, 1065 (1980).
14. R. Johnson and D. Johnson (unpublished).
15. T. J. Plona, App. Phys. Lett. *36*, 259 (1980).
16. T. J. Plona and D. L. Johnson, in *1980 Ultrasonic Symposium Proceedings*, edited by J. deKlerk and B. R. McAvoy (IEEE, NY, NY 1980), pp. 868-872.
17. R. C. Chandler and D. L. Johnson, J. Appl. Phys. (to be published); R. C. Chandler, J. Acoust. Soc. Am. (to be published).
18. T. Tanaka, L. O. Hocker, and G. B. Benedek, J. Chem. Phys. *59*, 5151 (1973).
19. D. L. Johnson (unpublished).
20. P. G. DeGennes, Macromolecule *9*, 587 (1976).
21. J. C. Bacri, J. M. Courdille, J. Dumas and R. Rajaonarison, J. Physique *41*, L-369 (1980); J. C. Bacri and R. Rajaonarison, J. Physique *40*, L-5 (1979).
22. D. L. Johnson and T. J. Plona (unpublished).
23. I. Rudnick, *New Directions in Physical Acoustics*, Proceedings of the Enrico Fermi Summer School, Course LXIII (Academic, NY, 1976), p. 112.
24. K. A. Shapiro and I. Rudnick, Phys. Rev. *137*, A1383 (1965).
25. R. J. S. Brown, Geophysics *45*, 1269 (1980).
26. P. N. Sen, C. Scala and M. H. Cohen, Geophysics *46*, 781 (1981).
27. T. J. Plona, D. L. Johnson, and C. Scala (unpublished).
28. D. L. Johnson and P. N. Sen, Phys. Rev. B (to be published).

BOUNDS FOR THE EFFECTIVE CONDUCTIVITY OF RANDOM MEDIA

W. Kohler and G. C. Papanicolaou

Virginia Polytechnic Institute and State University and
Courant Institute, New York University

Abstract

We formulate the problem of calculating the effective conductivity of a random medium in a suitable manner. Then we obtain upper and lower bounds using variational principles. Some of the bounds depend on the random geometry of the medium while others (like the Hashin-Shtrikman bounds) do not.

1. Introduction

We shall describe briefly our work on the determination of bounds for the effective conductivity of a random medium. Some additional information is contained in [1].

Bounds for effective parameters received a lot of attention in the past twenty or so years primarily because of the work of Hashin and Shtrikman [2]. In [3]-[7] extensions, improvements and new results are presented, always however within the framework of variational principles. Here we shall also stay within this framework. Our aim is (i) to remove the difficulties associated with the definition of effective parameters over finite samples of materials; (ii) derive geometry dependent bounds from multiple scattering expansions; and (iii) derive geometry independent bounds also from multiple scattering expansions.

Recently Bergmann [8] has introduced another method for deriving bounds that does not rely on variational principles. It works for two component media, relies on analytic continuation and can handle, for example, frequency dependent (hence complex-valued) effective dielectric constants (or conductivity). Milton [9] has used Bergman's method and has obtained further results. In [10] we give a mathematical formulation of Bergman's method similar to the one we employ here and we obtain a number of properties of the effective conductivity (dielectric constant) as a complex-valued function of a complex variable (the ratio of the conductivities of the two components).

2. Definition of effective conductivity

Let (Ω, Σ, P) be a probability space with $\omega \in \Omega$ labeling the realizations of the microscopic conductivity $a(x,\omega)$ which is a given stationary random function such that $0 < \alpha \leq a(x,\omega) \leq \beta < \infty$ for all $x \in \mathbb{R}^3$ and $\omega \in \Omega$. Stationary means that the joint distribution of $a(x_1,\omega), a(x_2,\omega), \ldots, a(x_N,\omega)$ for any set of points $x_1, x_2, \ldots, x_N$ is the same as that of $a(x_1+h,\omega), \ldots, a(x_N+h,\omega)$ for any $h \in \mathbb{R}^3$.

We denote by $G(x,\omega)$ and $F(x,\omega)$ the negative temperature gradient and the heat flux respectively at a point $x \in \mathbb{R}^3$ for the realization ω of the medium. The vector fields F and G are defined as follows. They are stationary random fields such that

$$\int_\Omega |G(x,\omega)|^2 P(d\omega) < \infty, \quad \int_\Omega |F(x,\omega)|^2 P(d\omega) < \infty$$

and

(2.1) $\quad F(x,\omega) = a(x,\omega) G(x,\omega)$

(2.2) $\quad \nabla \times G(x,\omega) = 0$

(2.3) $\quad \nabla \cdot F(x,\omega) = 0$

(2.4) $\quad \int_\Omega G(x,\omega) P(d\omega) = e$.

Here e is a fixed unit vector in $\mathbb{R}^3$. The effective conductivity $a^* = a^*(a,e,g)$ is defined by

(2.5) $\quad a^*(a,e,g) = \int_\Omega F(x,\omega) \cdot g \, P(d\omega)$

where g is another unit vector in $\mathbb{R}^3$. One can verify easily [1] that (2.1)-(2.4) has a unique solution and that a* is well defined.

The definition of a* is of course what one intuitively calls effective conductivity. The tensor $a^*(a,f,g)$ is the average heat flux in the direction g when unit average negative temperature gradient is prescribed in the direction e. The conducting material occupies an infinite region and its properties are statistically homogeneous. One can show [1] that the conductivity of a finite sample of material tends to a* for almost all realizations ω as the volume of the sample tends to infinity. This is a spatial ergodic theorem that justifies the usual operational definition of conductivity.

Working with the canonical problem (2.1)-(2.4) is advantageous because properties of a* can be obtained without the need to take limits, such as infinite volume limits.

To simplify exposition we shall assume in most of the sequel that a* is isotropic, i.e. it is independent of e and g.

Our objective is to get bounds for a* under various hypotheses about the values of the microscopic conductivity a and their statistical distribution. For this reason consider the following well-known variational principles.

The Dirichlet or energy principle gives

(2.6)
$$a^* = \min_{\substack{\nabla \times G = 0 \\ \langle G \rangle = e}} \langle aG \cdot G \rangle$$

where G runs over all stationary square integrable random fields that are curl-free and $\langle G \rangle$ denotes average

$$\langle G \rangle = \int_\Omega G(x,\omega) P(d\omega) \, .$$

The dual variational principle or Thomson's principle gives

(2.7)
$$\frac{1}{a^*} = \min_{\substack{\nabla \cdot F = 0 \\ \langle F \rangle = e}} \langle \frac{1}{a} F \cdot F \rangle \, .$$

The solution pair G and $(a^*)^{-1} F$ of (2.1)-(2.4) realizes the minimum in (2.6) and (2.7) respectively. Note that (2.5), (2.6) and (2.7) are all compatible since (in the isotropic case for simplicity)

$$a^* = \langle F \cdot e \rangle = \langle aG \cdot e \rangle = \langle aG \cdot G \rangle$$

because

$$\langle aG \cdot \tilde{G} \rangle = 0$$

for any test field $\tilde{G}$ with $\langle G \rangle = 0$ and $\nabla \times \tilde{G} = 0$. This follows from (2.3). We also have that

$$\langle \frac{1}{a}(a^*)^{-1} F \cdot (a^*)^{-1} F \rangle = (a^*)^{-2} \langle \frac{1}{a} F \cdot F \rangle$$
$$= (a^*)^{-2} \langle aG \cdot G \rangle$$
$$= (a^*)^{-2} \langle F \cdot e \rangle = (a^*)^{-1} \, .$$

The problem is then to find suitable trial fields G and F to insert in (2.6) and (2.7). For example, choosing $G = e$ and $F = e$ yields the very well known bounds

(2.8) $$(\langle \tfrac{1}{a} \rangle)^{-1} \leq a^* \leq \langle a \rangle .$$

These bounds are achieved by parallel plate 2-component materials.

Two questions arise in connection with bounds now:
(i) What are the best bounds one can obtain for a^* when the material is known to have specific structure such as spherical inclusions in a homogeneous matrix? It is assumed that the distribution functions of the spheres are known here.
(ii) What are the best bounds one can obtain for a^* when the geometrical structure of the composite material is very complex, for example a two-component material with neither of the components being a host or inclusion material?

Clearly in the second question we are asking for bounds that are somehow geometry-independent. The difference between the two types of bounds is quite analogous to the difference between the various kinds of approximate formulas for a^* that are commonly obtained. For example, the Claussius-Mosotti formulas [11] are results responding to questions of the first kind while the effective medium theories [12] (cf. also [13]) give results responding to questions of the second kind.

We close this section by deriving in a very simple manner two geometry independent results that are quite useful. The first is due to Keller [14] for a two-component periodic structure. It was generalized by Mendelshon [15]. It is proved here in full generality. The second is an inequality which generalizes Keller's result and was obtained by Schulgasser [16]. We prove it here by a different method in full generality.

<u>Theorem 1</u>
Let e_1 and e_2 be the orthogonal unit vectors in $\mathbb{R}^2$ such that e_2 is obtained from e_1 by a 90° rotation. Let $a^*(a, e_j, e_j) = a^*(a, e_j)$. Then

(2.9) $$\frac{1}{a^*(\tfrac{1}{a}, e_2)} = a^*(a, e_1)$$

<u>Remark</u>. In the isotropic case (2.9) reads

(2.10) $$\frac{1}{a^*(\tfrac{1}{a})} = a^*(a) .$$

Since for any positive constant γ, $a^*(\gamma a) = \gamma a^*(a)$, when a takes only the values $a = a_1$ or a_2 then the relation (2.10) can be written

$$1 = a^*(\frac{1}{a_1}, \frac{1}{a_2}) a^*(a_1, a_2)$$

or

(2.11) $$a_1 a_2 = a^*(a_2, a_1) a^*(a_1, a_2) .$$

The form (2.10) is more appropriate since it holds in general, not just a two-component medium.

Theorem 2

Let e_1, e_2 and e_3 be an orthonormal triplet of vectors in $\mathbb{R}^3$ so that $e_1 \times e_2 = e_3$, $e_2 \times e_3 = e_1$, $e_3 \times e_1 = e_2$. Then

(2.12) $$\frac{1}{a^*(\frac{1}{a}, e_i \times e_j)} \leq a^*(a, e_j) , \qquad i,j = 1,2,3, \quad i \neq j .$$

Remark: In the isotropic, two-component case (2.12) reduces to

(2.13) $$a_1 a_2 \leq a^*(a_2, a_1) a^*(a_1, a_2)$$

and as Schulgasser [16] observed, strict inequality holds for the coated sphere medium (used by Bruggeman) that realizes the upper and lower Hashin-Shtrikman bounds.

Proof of Theorem 1

From (2.7)

(2.14) $$\frac{1}{a^*(\frac{1}{a}, e_2)} = \min_{\substack{\nabla \cdot F = 0 \\ \langle F \rangle = 0}} \langle a(e_2 + F) \cdot (e_2 + F) \rangle .$$

Let S be the 90° rotation map that takes e_1 into e_2, i.e. $Se_1 = e_2$. Let G be a two-dimensional, zero mean stationary random field with $\nabla \times G = 0$, i.e. $\partial G_1 / \partial x_2 = \partial G_2 / \partial x_1$. Let

(2.15) $$\tilde{F} = SG .$$

Then clearly $\tilde{F}$ has mean zero is stationary and $\nabla \cdot \tilde{F} = 0$. Moreover

(2.16) $\langle a(e_2+\tilde{F}) \cdot (e_2+\tilde{F}) \rangle = \langle aS(e_1+G) \cdot S(e_1+G) \rangle$

$$= \langle a(e_1+G) \cdot (e_1+G) \rangle .$$

Since the mapping (2.15) is one-one and onto we conclude that taking minima in (2.16) yields (2.9) in view of (2.14) and the analogous expression from (2.6).

Proof of Theorem 2

The proof is similar to the one above. The field $e_i \times e_j + e_i \times G$, where G is stationary, zero mean and $\nabla \times G = 0$, is an admissible field in the variational principle (2.7), i.e.

$$\frac{1}{a^*(\frac{1}{a}, e_i \times e_j)} = \min_{\substack{\nabla \cdot F = 0 \\ <F> = 0}} <a(e_i \times e_j + F) \cdot (e_i \times e_j + F)> .$$

But the class of divergence-free fields is not spanned by fields of the form $e_i \times G$, $\nabla \times G = 0$ as was the case in dimension 2. Thus

$$\frac{1}{a^*(\frac{1}{a}, e_i \times e_j)} \leq \min_{\substack{\nabla \times G = 0 \\ <G> = 0}} <a(e_i \times e_j + e_i \times G) \cdot (e_i \times e_j + e_i \times G)>$$

$$= \min_{\substack{\nabla \times G = 0 \\ <G> = 0}} <a[e_i \times (e_j + G)] \cdot [e_i \times (e_j + G)]>$$

$$= \min_{\substack{\nabla \times G = 0 \\ <G> = 0}} \{<a(e_j + G) \cdot (e_j + G)> - <a[e_i \cdot (e_j + G)]^2>\}$$

$$\leq \min_{\substack{\nabla \times G = 0 \\ <G> = 0}} <a(e_j + G) \cdot (e_j + G)>$$

$$= a^*(a, e_j) ,$$

as was to be shown.

3. Geometry-independent bounds

The difficulty in applying the variational principles (2.6) and (2.7) comes of course from the fact that the trial fields must be curl-free and divergence-free, respectively. To overcome this difficulty, i.e. in order to be able to construct many interesting admissible fields, we look again at the Euler equations for (2.6) which are (2.1)-(2.4).

We introduce, as did Hashin and Shtrikman [2], a reference medium with constant conductivity a_o and rewrite (2.3) as

(3.1) $\quad \nabla \cdot (a_o G) + \nabla \cdot (\delta a G) = 0, \qquad\qquad \nabla \times G = 0,$

where

(3.2) $\qquad\qquad \delta a = a(x,\omega) - a_o .$

Let us define formally an operator Γ by

(3.3) $\qquad\qquad \Gamma G = -\nabla(-\Delta)^{-1}(\nabla \cdot G) .$

Lemma 1

The operator Γ given formally by (3.3) is a well-defined projection operator on the Hilbert space

(3.4) $\qquad H = \left\{ G \mid G \text{ is stationary vector field, } \langle G \rangle = 0, \langle G \cdot G \rangle < \infty \right\}$

into the subspace of curl free fields in H. In particular, Γ satisfies

(3.5) $\qquad \Gamma^2 = \Gamma .$

Lemma 2

If G in H is in addition isotropic,

$$\langle G_i(x+y) G_j(y) \rangle = R_{ij}(|x|) \qquad i,j = 1,2,3,$$

i.e., the correlation function depends on $|x|$ only, then

(3.6) $\qquad \langle \Gamma G \cdot G \rangle = \frac{1}{3} \langle G \cdot G \rangle .$

The proof of both lemmas is quite simple and we shall not pause to give it (cf. [1]).

With the help of the projection operator Γ we can write (3.1) in the form

(3.7) $\qquad G + \Gamma(a_o^{-1} \delta a G - \langle a_o^{-1} \delta a G \rangle) = e .$

We observe that in fact (3.7) summarizes at once all of (2.1)-(2.4). However, (3.7) is not the Euler equation of a variational problem as it stands because the operator $\Gamma(a_o^{-1} \delta a \cdot)$ is not selfadjoint in H even though the projection operator Γ is. One way to transform (3.7) to the Euler equation of a variational problem is to let the unknown field be $a_o^{-1} \delta a G$, rather than G. Let

(3.8) $\qquad\qquad P = a_o^{-1} \delta a G$

so that (3.7) becomes

(3.9) $\quad a_o(\delta a)^{-1} P + \Gamma(P-\langle P\rangle) = e$

Clearly (3.9) is equivalent to (3.7) and by analogy with electrostatics we call P the polarization field.

Equation (3.9) is the Euler equation of the quadratic functional

(3.10) $\quad U(P) = \frac{1}{2} \langle a_o(\delta a)^{-1} P \cdot P\rangle + \frac{1}{2} \langle \Gamma(P-\langle P\rangle) \cdot (P-\langle P\rangle)\rangle - \langle e \cdot P\rangle$

on $\tilde{H} = H +$ constants. Note in particular that the functional U is defined over all of $\tilde{H}$ without constraints. One verifies easily that when $a_o < a(x,\omega)$ for all x and ω, so that $\delta a > 0$, then $U(P)$ takes its minimum at $P = P$ of (3.8) and

(3.11) $\quad U(P) = \frac{1}{2}\left(1 - \frac{a^*}{a_o}\right)$.

Thus

(3.12) $\quad \frac{1}{2}\left(1 - \frac{a^*}{a_o}\right) \leq U(P)$ when $\delta a > 0$

where P is an arbitrary field in $\tilde{H}$. The bound (3.12) can be rewritten as

(3.13) $\quad a_o(1 - 2U(P)) \leq a^*$, when $\delta a > 0$.

Note that a_o is also arbitrary provided $0 < a_o < a(x,\omega)$.

When $\delta a < 0$, i.e. $a(x,\omega) < a_o < \infty$, then the quadratic form (3.10) is negative definite in $\tilde{H}$. This is clear because $a_o(\delta a)^{-1} < -1$ and since Γ is a projection $\|\Gamma\| \leq 1$. Thus

(3.14) $\quad a^* \leq a_o(1 - 2U(P))$, when $\delta a < 0$.

The next step is the selection of a trial field P for (3.13) and (3.14). Obviously we want a trial field as close to P as possible. Let us look at (3.9) which we rewrite as

(3.15) $\quad P = \frac{\delta a}{a_o} [e - \Gamma(P-\langle P\rangle)]$.

There are several ways in which we can construct trial fields close to P depending on what we know about the medium. We single out 3 cases.

Case 1. $|\delta a|$ is small for some a_o, i.e. the random medium is nearly uniform.

In this case we can expand P in (3.15) in power series in $|\delta a|$ and obtain trial fields by truncating the series. For example,

$$P = \frac{\delta a}{a_o} e - \frac{\delta a}{a_o} \Gamma \left(\frac{\delta a}{a_o} e - \langle \frac{\delta a}{a_o} e \rangle \right)$$

is an admissible field, The calculations are not particularly interesting here although with the help of the analytic continuation method (for two-component media, cf. [8]-[10]), they can become significant.

<u>Case 2</u>. δa is not small for any a_o but the medium has a host-inclusion geometry and the multiple scattering formalism can be used. This is discussed further in the next section.

<u>Case 3</u>. δa is not small for any a_o and the medium is not of the simple host-inclusion type. We suppose however that we know that the conductivity $a(x,\omega)$ is statistically isotropic.

We shall pursue here this last case because it is geometry-independent and leads to interesting results.

We assume from now on that $a(x,\omega)$ is statistically isotropic.

We note first that the operator Γ on H defined by (3.3) can be written in the form

(3.16) $$\Gamma = \Gamma_0 + \Gamma_1$$

where

(3.17) $$\Gamma_0 = \frac{1}{3} I \quad \text{and} \quad \Gamma_1 = \Gamma - \frac{1}{3} I .$$

Because of lemma 2 and (3.6) we see that for any <u>isotropic</u> fields G_1 and G_2 in H

(3.18) $$\langle \Gamma_1 G_1 \cdot G_2 \rangle = 0 .$$

To verify (3.18) one has to note that (3.6) implies that

(3.19) $$\langle \Gamma G_1 \cdot G_2 \rangle = \frac{1}{3} \langle G_1 \cdot G_2 \rangle$$

for any G_1, G_2 in H which are isotropic.

We now return to (3.9) and rewrite it in the form

$$(3.20) \quad P - \langle P \rangle + \frac{\delta a}{a_0} \Gamma_0 (P - \langle P \rangle) + \frac{\delta a}{a_0} \Gamma_1 (P - \langle P \rangle) = \frac{\delta a}{a_0} e - \langle P \rangle$$

or

$$\left[I + \left(I + \frac{\delta a}{a_0} \Gamma_0 \right)^{-1} \frac{\delta a}{a_0} \Gamma_1 \right] (P - \langle P \rangle) = \left(I + \frac{\delta a}{a_0} \Gamma_0 \right)^{-1} \left[\frac{\delta a}{a_0} e - \langle P \rangle \right] .$$

Let

$$(3.21) \quad L = -\left(I + \frac{\delta a}{a_0} \Gamma_0 \right)^{-1} \frac{\delta a}{a_0} \Gamma_1 .$$

Then

$$(3.22) \quad P - \langle P \rangle = \sum_{n=0}^{\infty} L^n \left[\left(I + \frac{\delta a}{a_0} \Gamma_0 \right)^{-1} \left(\frac{\delta a}{a_0} e - \langle P \rangle \right) \right] .$$

We now choose as a trial field for (3.13) or (3.14) the simplest P compatible with (3.22), namely

$$(3.23) \quad P - \langle P \rangle = \left(I + \frac{\delta a}{a_0} \Gamma_0 \right)^{-1} \left(\frac{\delta a}{a_0} e - \langle P \rangle \right) .$$

We thus neglect all terms in (3.22) that involve Γ_1. Taking averages in (3.23), noting that $\Gamma_0 = \frac{1}{3} I$ and rearranging we find that

$$(3.24) \quad \langle P \rangle = \left\{ \left\langle \left(I + \frac{\delta a}{3 a_0} \right)^{-1} \right\rangle \right\}^{-1} \left\langle \left(I + \frac{\delta a}{3 a_0} \right)^{-1} \frac{\delta a}{a_0} \right\rangle e .$$

We evaluate next the functional U of (3.10) on the trial field P of (3.23). This gives

$$(3.25) \quad U(P) = -\frac{1}{2} \langle P \rangle \cdot e .$$

To verify (3.25) we note that since a is isotropic the Γ_1 term $(\Gamma = \Gamma_0 + \Gamma_1)$ in (3.10) drops out and that (3.23) can be rewritten in the form

$$3.26) \quad \frac{a_0}{\delta a} P + \Gamma_0 (P - \langle P \rangle) = e ,$$

which simplifies computations.

Inserting (3.24) into (3.25) and the result into (3.13) and (3.14) we obtain the bounds

(3.27) $\quad a_0\left\{1 + \left[<(1+\frac{\delta a}{3a_0})^{-1}>\right]^{-1} <(1+\frac{\delta a}{3a_0})^{-1}\frac{\delta a}{a_0}>\right\} \leq a^*$

when $\delta a > 0$

and

(3.28) $\quad a^* \leq a_0\left\{1 + \left[<(<1+\frac{\delta a}{3a_0})^{-1}>\right]^{-1} <(1+\frac{\delta a}{3a_0})^{-1}\frac{\delta a}{a_0}>\right\}$

when $\delta a < 0$.

Now suppose that $a(x,\omega)$ takes only two values (or any finite number) a_1 and a_2 with $0 < a_1 < a_2 < \infty$. Then the inequality (3.27) is optimized by taking $a_0 = a_1$ and (3.28) by taking $a_0 = a_2$. A simple calculation yields now the bounds

(3.29) $\quad a_1 + \dfrac{\rho_2}{\dfrac{1}{a_2-a_1} + \dfrac{\rho_1}{3a_1}} \leq a^* \leq a_2 + \dfrac{\rho_1}{\dfrac{1}{a_1-a_2} + \dfrac{\rho_2}{3a_2}}$

which are the bounds of Hashin and Shtrikman [2]. In (3.29) we have set

(3.30) $\quad \rho_j = P\{a(x) = a_j\}, \quad j=1,2$.

To improve on the bounds (3.27)-(3.28), or (3.29) for the two component medium, one has to use in (3.13)-(3.14) a better trial field P. From (3.22) we see that this should be

(3.31) $\quad P - <P> = (I + \frac{\delta a}{a_0}\Gamma_0)^{-1}(\frac{\delta a}{a_0}e - <P>)$

$\quad - (I + \frac{\delta a}{a_0}\Gamma_0)^{-1}\frac{\delta a}{a_0}\Gamma_1\left[(I + \frac{\delta a}{a_0}\Gamma_0)^{-1}(\frac{\delta a}{a_0}e - <P>)\right]$.

However this is not quite appropriate because it does not satisfy

(3.32) $$\langle \frac{a_0}{\delta a} P \rangle = e$$

as the exact field P does from (3.8) and (2.4). To get the appropriate higher order trial field $\tilde{P}$ we look again at (3.15). From (3.9) we can write formally

(3.33) $$P - \langle P \rangle = \left(I + \frac{\delta a}{a_0} \Gamma\right)^{-1} \left(\frac{\delta a}{a_0} e - \langle P \rangle\right).$$

Using this in (3.15) we obtain the iterated form

(3.34) $$P - \langle P \rangle = \frac{\delta a}{a_0} e - \langle P \rangle - \frac{\delta a}{a_0} \Gamma \left[\left(I + \frac{\delta a}{a_0}\Gamma\right)^{-1} \left(\frac{\delta a}{a_0} e - \langle P \rangle\right)\right].$$

Now we take as trial field

(3.35) $$\tilde{P} - \langle P \rangle = \frac{\delta a}{a_0} e - \langle P \rangle$$
$$- \frac{\delta a}{a_0} \Gamma \left[\left(I + \frac{\delta a}{a_0}\Gamma_0\right)^{-1}\left(\frac{\delta a}{a_0} e - \langle P \rangle\right)\right].$$

This can be rewritten in the form

(3.36) $$\tilde{P} - \langle P \rangle = \left(I + \frac{\delta a}{a_0}\Gamma_0\right)^{-1}\left(\frac{\delta a}{a_0} e - \langle P \rangle\right)$$
$$- \frac{\delta a}{a_0}\Gamma_1\left[\left(I + \frac{\delta a}{a_0}\Gamma_0\right)^{-1}\left(\frac{\delta a}{a_0} e - \langle P \rangle\right)\right]$$

which shows the difference with (3.31).

The trial field (3.36) is now used in (3.13)-(3.14). Of course the terms involving the operator Γ_1 do not drop out now and it is necessary to have information about correlation functions in order to calculate explicitly the resulting bounds. The calculations are given in [1].

One can also work directly with the variational principles (2.6)-(2.7) by using trial fields based on multiple scattering expansion, for example from (3.7) (cf. [7]). If is more convenient however to work with (3.13)-(3.14) and the above method for constructing trial fields.

4. Geometry dependent bounds

Let $\{y_j\}$ be a stationary distribution of points in $\mathbb{R}^3$. A realization of the points will be denoted by $\omega = \{y_j\}$. We now define the conductivity $a(x,\omega)$ by

$$(4.1) \qquad a(x,\omega) = \begin{cases} a_2, & |x-y_j| \leq \delta \text{ for some } j \\ a_1, & |x-y_j| > \delta \text{ for all } j \end{cases}.$$

We assume that the spheres of radius δ centered at $\{y_j\}$ do not overlap. We denote by ρ the volume fraction occupied by the spheres

$$(4.2) \qquad \rho = P\{a(x) = a_2\} = \tfrac{4}{3}\pi\delta^3\alpha$$

where α is the average number of sphere centers per unit volume. The ρ here corresponds to ρ_2 of the previous section.

We wish to find upper and lower bounds for the effective conductivity a^* which when ρ is small are much closer than the Hashin-Shtrikman bounds (3.29). The improvement must come of course from the knowledge we have now of the random geometry: conducting spheres distributed in a uniform medium. Unfortunately the two and three point functions of the point process $\{y_j\}$ that are necessary for our calculations are not known for realistic cases that allow dense packing with volume exclusion. To render our formulas explicit we shall use the heuristic "well-stirred" approximation which we now describe.

Let $f(x), g(x)$ and $h(x)$ be smooth functions of compact support in $\mathbb{R}^3$. We calculate

$$(4.3) \quad \langle \sum_j f(y_j) \sum_k g(y_k) \sum_\ell h(y_\ell) \rangle = \alpha \int f(x) g(x) h(x) p_1(x) dx$$

$$+ \alpha^2 \iint f(x) g(x) h(y) p_2(x,y) dxdy + \alpha^2 \iint f(x) h(x) g(y) p_2(x,y) dxdy$$

$$+ \alpha^2 \iint g(x) h(x) f(y) p_2(x,y) dxdy$$

$$+ \alpha^3 \iiint f(x) g(y) h(z) p_3(x,y,z) dxdydz .$$

Here α is the average number of points $\{y_j\}$ per unit volume, $p_1(x)$ is identically equal to one (by stationarity), p_2 and p_3 are the two and three point functions respectively. By stationarity, $p_2(x,y) = p_2(0,x-y)$ and $p_3(x,y,z) = p_3(0,y-x,z-x)$. The well-stirred approximation consists in taking

$$(4.4) \quad p_2(z) = \begin{cases} 0, & |z| \le \delta \\ 1, & |z| > \delta \end{cases}, \quad p_3(y,z) = \begin{cases} 0, & |y| \le 1 \\ 0, & |z| \le 1 \\ 0, & |y-z| \le 1 \\ 1, & \text{all other cases.} \end{cases}$$

There are less naive models for the function p_2 (cf. [19]) which are still not entirely rigorous but have been used with success. For the function p_3 the situation is quite complicated and little seems to be known about it for realistic situations (i.e. that allow close packing).

Now our approach is to use the classical variational principles (2.6) and (2.7) directly. Trial fields are constructed by multiple scattering expansions [7,13] as follows.

Let

$$(4.5) \quad v_j(x) = \begin{cases} 1, & |x-y_j| \le \delta \\ 0, & |x-y_j| > \delta \end{cases}.$$

We suppose for the moment that j runs from 1 to $N < \infty$ and write (3.1) in the form

$$(4.6) \quad \nabla \cdot (a_0 G) + \sum_{j=1}^{N} \nabla \cdot \left[(a_2-a_1)v_j(x) + \frac{a_1-a_0}{N} \right) G \right] = 0 .$$

Now let

$$L_0 = \nabla \cdot (a_0 \cdot), \qquad M_0 = \sum_{j=1}^{N} v_j^{(0)}$$

$$(4.7) \quad v_j^{(0)} = \nabla \cdot \left[((a_2-a_1)v_j + \frac{a_1-a_0}{N}) \cdot \right]$$

so that (4.6) becomes

(4.8) $$(L_0 + M_0)G = 0 \ .$$

Define further the operator T_0 by

(4.9) $$T_0 = -(L_0 + M_0)^{-1}M_0 \ .$$

If we set

(4.10) $$G = (I + T_0)(I + <T_0>)^{-1}e \ ,$$

we can verify that this field G satisfies (4.6); in fact (2.1)-(2.4), if in (4.8) and the definition of T_0 we include $\nabla \times G = 0$.

From (4.7) and (4.9)

(4.11) $$\begin{aligned} T_0 &= -\left(L_0 + \sum_k V_k^{(0)}\right)^{-1} \sum_j V_j^{(0)} \\ &= -\sum_j \left(L_0 + V_j^{(0)} + \sum_{k \neq j} V_k^{(0)}\right)^{-1} V_j^{(0)} \\ &= -\sum_j \left[I + (L_0+V_j^{(0)})^{-1} \sum_{k \neq j} V_k^{(0)}\right]^{-1} (L_0+V_j^{(0)})V_j^{(0)} \\ &= \sum_j \left[I - \sum_{k \neq j} T_{jk}^{(0)}\right]^{-1} T_j^{(0)} \ . \end{aligned}$$

Here

(4.12) $$T_j^{(0)} = -(L_0 + V_j^{(0)})V_j^{(0)} \ , \quad T_{jk}^{(0)} = (L_0 + V_j^{(0)})^{-1}V_k^{(0)} \ .$$

The form (4.11) of the "scattering" operator T_0 is the multiple scattering form when the inverse operator in the last line of (4.11) is expanded further. If we approximate T_0 by

(4.13) $$T_0 \sim \sum_j T_j^{(0)} \ ,$$

then a trial field can be constructed by taking

(4.14) $$G = \left(I + \sum_j T_j^{(0)}\right)\left(I + <\sum_j T_j^{(0)}>\right)^{-1} e \ .$$

Thus we must find the explicit form of $T_j^{(0)}$ on uniform fields. This is an elementary computation that yields

$$(4.15) \qquad (T_j^{(0)}\tilde{e})(x) = -\gamma\left[v_j\tilde{e} + \delta^3(1-v_j)\nabla_x\left(\frac{(x-y_j)\cdot\tilde{e}}{|x-y_j|^3}\right)\right]$$

where $\tilde{e}$ is some uniform field and

$$(4.16) \qquad \gamma = \frac{a_2-a_1}{2a_0+a_2-a_1}.$$

Here we have already passed to the limit $N \to \infty$ (cf. (4.7) and (4.12)) and, in order that $\langle\sum_j T_j\rangle$ exist we assume that the field $T_j^{(0)}\tilde{e}$ is cut off smoothly when $|x-y_j| > 10\delta$ say. Because of this $\langle\sum T_j\rangle = 0$ and hence the trial field takes the form

$$G = e + \sum_j T_j^{(0)} e$$

with $T_j^{(0)}e$ defined by (4.15).

We insert this G into (2.6) and perform the averages as indicated in (4.3) with the use of the well-stirred approximation (4.4). This is a very lengthy computation which we shall not present. At the end of the computation the constant a_0 is chosen to optimize the bound. We simply state the result of our computations

$$(4.17) \qquad \frac{a^*}{a_1} \leq 1 + \rho(K-1) - \frac{(K-1)^2\rho(1-\ )^2}{3-(2-\kappa)\rho+(K-1)(1-2)(1-\xi)\rho+\eta\rho^2)}.$$

Here

$$(4.18) \qquad K = \frac{\dot{a}_2}{a_1}$$

and κ, ξ and η are constants that stand for integrals of dipole-like functions over multidimensional domains with excluded regions arising from (4.4). To three decimal point accuracy

$$(4.19) \qquad \kappa = -.832, \qquad \xi = .176$$

while a crude bound for η is

$$(4.20) \qquad |\eta| \leq 101.43.$$

For $\rho \ll 1$, the bound (4.17) expanded to $O(\rho^2)$ is

(4.21) $\quad \dfrac{a^*}{a_1} \leq 1 + 3\left(\dfrac{K-1}{K+2}\right)\rho + [3.168 + .352(K-1)]\left(\dfrac{K-1}{K+2}\right)^2 \rho^2 + \ldots$.

An entirely analogous computation can be carried out for a lower bound based on the trial field

(4.22) $\quad F = (1+3\rho\gamma)^{-1}\left\{ e - \gamma \sum_j \left[(1-v_j)\delta^3 \nabla_x \left(\dfrac{(x-y_j)\cdot e}{|x-y_j|^3} \right) - 2v_j e \right] \right\}$

which is admissible for (2.7). Here the smooth cutoff assumption explained below (4.16) is in effect and γ is given by (4.16).

Again the calculation of $\langle a^{-1} F \cdot F \rangle$ for the F of (4.22) is very lengthy. After it is completed one must choose a_0 to optimize the bound that results from (2.7). We do not give the result in general since it is a very complicated expression. We give only the expanded to $O(\rho^2)$ form analogous to (4.21) which is

(4.23) $\quad \dfrac{a^*}{a_1} \geq 1 + 3\left(\dfrac{K-1}{K-2}\right)\rho + \left[2.832 + .352\,\dfrac{K-1}{K}\right]\left(\dfrac{K-1}{K+2}\right)^2 \rho^2 + \ldots$.

We must now compare the bounds (4.21) and (4.23) with the expanded form of the Hashin-Shtrikman bounds (3.29). We rewrite (3.29) with the notation of (4.21)-(4.23):

$$1 + \dfrac{\rho}{\dfrac{1}{K-1} + \dfrac{1-\rho}{3}} \leq \dfrac{a^*}{a_1} \leq K + \dfrac{1-\rho}{\dfrac{1}{1-K} + \dfrac{\rho}{3K}} .$$

Expanding the lower bound for small ρ we get

(4.24) $\quad 1 + 3\left(\dfrac{K-1}{K+2}\right)\rho + 3\rho^2\left(\dfrac{K-1}{K+2}\right)^2 \rho + \ldots \leq \dfrac{a^*}{a_1}$.

We see therefore that the upper bound (4.21) is above the lower Hashin-Shtrikman bound but close to it while the lower bound (4.23) is actually below (4.24) but close again. Of course it is impossible to realize media with given ρ and a^* below the one given by (3.29). Our result

(4.23) reflects the particular nature of the well stirred approximation (4.4) and its limitations. We could have expressed our results (4.21), (4.23) in terms of constants that involve integrals of dipole fields, p_2 and p_3. Then the lower bound would surely have been above the Hashin-Shtrikman lower bound.

We feel that the calculations with the well-stirred approximation (4.4) are useful in showing that geometry-dependent bounds can be derived that are sharp to order ρ as $\rho \downarrow 0$ (i.e. the order ρ term in (4.21) and (4.23) are the same). This cannot be the case with geometry-independent bounds.

It would be interesting to find the form of the bounds (4.21) and (4.23) when the Percus-Yevick two-point function [19],[21] and its analogous three-point function are used.

Let us also compare the above bounds with results from selfconsistent calculations. We compare with formula (8.23) of [13]

$$(4.25) \qquad a^* = \frac{\langle a \rangle - \rho a_2 \left(\frac{a_2 - a_1}{3a^* - a_2 - a_1} \right)}{1 - \rho \frac{a_2 - a_1}{3a^* + a_2 - a_1}}$$

which for small ρ is

$$(4.26) \qquad \frac{a^*}{a_1} = 1 + 3\left(\frac{K-1}{K+2}\right)\rho + 9\left(\frac{K-1}{K+2}\right)^2 \rho^2 + \ldots \ .$$

We see that the upper bound (4.21) is below (4.26) except when K is large ($K \geq 1$ here). Since (4.25) is expected to be a good approximation for the spherical inclusion composite conductor under consideration, we see again that the well-stirred approximation is quite crude when used for bounds.

The effective medium theory [11] gives for a^* the solution of

$$(4.27) \qquad \frac{a^* - a_1}{2a^* + a_1} (1-\rho) + \frac{a^* - a_2}{2a^* + a_2} \rho = 0$$

which for ρ small becomes

$$\tag{4.28} \frac{a^*}{a_1} = 1 + 3\frac{K-1}{K+2}\rho + 9\left(\frac{K-1}{K+2}\right)^2 \frac{1}{K+2}\rho^2 + \ldots .$$

The Claussius-Mosotti formula [11] gives

$$\frac{a^*}{a_1} = 1 + 3\frac{K-1}{K+2}\rho + 3\left(\frac{K-1}{K+2}\right)^2 \rho^2 + \ldots$$

and Böttcher's result [17] is

$$\frac{a^*}{a_1} = 1 + 3\frac{K-1}{K+2}\rho + \rho^2\left[\frac{3(K-1)^2}{(K+2)^2} + 6\left(\frac{K-1}{K+2}\right)^3\right] + \ldots$$

All these formulas have unknown regions of validity but for small ρ, say $\rho = .1$ or $\rho = .15$, they are not too bad when compared with experimental data [20].

We conclude by noting again that computation of bounds by multiple scattering when sophisticated two and three point functions are used (and the resulting integrals are computed numerically) would produce inequalities analogous to (4.21)-(4.23), or their unexpanded form, which could be very useful.

This work was supported by the Army Research Office under Grant #DAAG29-78-G-0177.and #DAAG29-79-C-0085.

REFERENCES

[1] W. Kohler and G. Papanicolaou, Upper and lower bounds for effective conductivities, to appear.

[2] Z. Hashin and S. Shtrikman, A variational approach to the theory of effective magnetic permeability of multiphase materials, J. Applied Phys. 33 (1962), pp. 3125-3131.

[3] L.J. Walpole, On bounds for the overall elastic moduli of inhomogeneous systems I, II, J. Mech. Phys. Solids, 14 (1966), pp. 151-162 and 289-301

[4] R. Hill, Elastic properties of reinforced solids: some theoretical principles, J. Mech. Phys. Solids, 11 (1963), pp. 357-372.

[5] M. Beran, Statistical Continuum Theories, Wiley, New York, 1968.

REFERENCES, continued

[6] J.R. Willis, Bounds and self-consistent estimates for the overall properties of anisotropic composites, J. Mech. Phys. Solids, 25 (1977), pp. 185-202.

[7] P.H. Dederichs and R. Zeller, Variational treatment of elastic constants of disordered materials, Z. Physik 259 (1973), pp. 103-116.

[8] D. Bergman, The dielectric constant of a composite material -- a problem in classical physics, Phys. Rep. C, 43 (1978), pp. 377-407.

[9] G. Milton, Bounds on the transport and optical properties of a two-component composite material, J. Appl. Phys., to appear.

[10] K. Golden and G. Papanicolaou, Bounds for effective parameters of heterogeneous media by analytic continuation, to appear.

[11] R. Landauer, Electrical conductivity in inhomogeneous media. AIP Conf. Proc. #40, J. Garland and D. Tanner, editors, Am. Inst. of Phys., New York, 1978.

[12] D.A.G. Bruggeman, Berechnung verschiedener physicalischer Konstanten von heterogenen Substanzen, Annalen der Physik 22 (1935), pp. 636-679.

[13] W. Kohler and G. Papanicolaou, Some applications of the coherent potential approximation, in Multiple Scattering and Waves in Random Media, edited by Chow, Kohler and Papanicolaou, North Holland, 1981.

[14] J.B. Keller, A. Theorem on the conductivity of a composite medium, J. Math. Phys. 5 (1964), pp. 548-549.

[15] K.S. Mendelshon, Effective conductivity of a two-phase material with cylindrical phase boundaries, J. Appl. Phys. 46 (1975), pp. 917-918.

[16] K. Schulgasser, On a phase interchange relationship for composite materials, J. Math. Phys. 17 (1976), p. 378.

[17] W.F. Brown, Jr., in Handbuch der Physik 17, Springer Verlag, Berlin, 1956.

[18] C.J.F. Böttcher, Theory of Electric Polarization, Elsevier, Amsterdam, 1973.

[19] G. Stell, G.N. Patey, J.S. Hoye, Dielectric constants of fluid models: statistical mechanical theory and its quantitative implementation, Adv. in Chem. Phys. 48, J. Wiley and Sons, New York, 1981.

[20] Z. Hashin, Theory of composite materials, in Mechanics of Composite Materials, edited by F.W.Wendt, H. Lebowitz and N. Perrone, Pergamon Press, 1970, pp. 201-242.

[21] J.K. Percus and G.J. Yevick, Hard-core insertion in the many-body problem, Phys. Rev. 136 (1964), pp. B290-B296.

STRUCTURAL DESIGN OPTIMIZATION, HOMOGENIZATION AND RELAXATION OF VARIATIONAL PROBLEMS

Robert V. Kohn
Courant Institute of Mathematical Sciences

Gilbert Strang
Massachusetts Institute of Technology

We propose to survey some relationships between three distinct areas of investigation:

I) The design of engineering structures to optimize a given performance criterion.
II) The determination of "optimal" bounds for the effective properties of a composite material, in terms of the volume fractions of the components.
III) The determination of the lower-semicontinuous hull of a non-convex variational problem.

A separate section is devoted to each area. In section 1 we survey briefly - and rather idiosyncratically - some problems that have been the focus of attention in the structural optimization literature, and what is known about them. Section 2 explains why finding sharp bounds for effective equations is essentially equivalent to solving a broad class of geometry optimization questions. This idea is not new to experts; it is implicit in recent work by Murat, Tartar, Olhoff, Cheng, Lurie, and others. We believe, however, that it has not been made sufficiently explicit elsewhere, so we have tried to explain it here. Section 3 describes a program, initiated recently by the authors, for using the relaxation of non-convex variational problems as a tool in the theory of optimal design. Both sections 2 and 3 include a number of explicit open problems.

We acknowledge here with gratitude a great debt to many colleagues who have shared their expertise and insight over a period of several years. Particular thanks go to John Taylor and Niels Olhoff regarding structural optimization, to Luc Tartar and George Papanicolaou regarding homogenization, and to Roger Temam regarding convex duality.

This work has been supported in part by NSF grants MCS-7919146 A01 (to R.V.K.) and MCS81-02371 (to G.S.), and also by Army research contract DAAG29-80-K0033 (G.S.).

§1: Structural Optimization

We discuss a number of different design optimization questions, always with this basic form: given a design problem in which the shape or composition of the structure is to some extent variable, but the loads it must withstand are fixed, choose a design that "performs optimally" subject to constraints on its cost.

A simple version of this problem - in the context of buckling of columns - was considered as early as 1770 by Lagrange [1]; a fundamental paper was written by Michell in 1904 on the optimal layout of trusses [2]. Most of the now vast literature, however, developed in the past 25 years - stimulated, it seems, largely by the aerospace industry, where weight is critical, and by the availability of computers for carrying out optimization algorithms.

Reviewing this extensive literature has become an almost hopeless task; for reviews of various parts the reader may look to [3,4,5,6], especially the article by Haug in [3]. Our more limited goal here is to summarize some of the themes running though this work. Following common practice, we restrict our attention to the linearized models of behavior: linear elasticity and linearly elastoplastic behavior. Most of the questions discussed herein have analogues for more realistic, nonlinear models, but it is difficult at this point to say much about the corresponding optimization problems.

There are many different criteria that one might use to gauge the efficiency of a structure in withstanding a given load; certain of these have dominated the recent literature. For problems of linear elastostatics, one typically seeks a design that minimizes

    1.1a) The work done by the load upon the structure (its "_compliance_" under the load);

or

    1.1b) The maximum stress produced by the load (taken pointwise, using a suitable norm upon symmetric tensors);

or

    1.1c) The maximum displacement produced by the load.

For elasto-plastic materials (the model of choice for ductile metals) a corresponding performance criterion is to maximize

    1.1d) The _limit multiplier_ of the load (by definition, the largest multiple of the load which the structure can withstand without developing global plastic flow).

In addition to these static performance criteria, there are many important problems relating to eigenvalues of linearly elastic structures, interpreted as buckling modes or as modes of vibration. One might attempt to maximize the lowest eigenvalue; or, in a vibration problem, to create a large gap in the spectrum near a particular frequency. For structures made from elastoplastic materials, dynamic optimization problems arise also in avoiding fatigue under cyclic loading conditions.

In many problems - particularly those of aerospace design - the "cost" of the structure may be interpreted as the total weight. One might seek either designs that perform optimally for fixed cost, or ones that minimize the cost for fixed strength.

What kind of "structures" do we have in mind? Perhaps the simplest problem conceptually is that of fully three dimensional <u>material distribution optimization</u>: the data are

1.2a) a domain $\Omega \subset \mathbb{R}^3$

1.2b) a finite collection of linearly elastic materials, each homogeneous, characterized by the stress-strain laws $\sigma = A^{(r)}(\varepsilon)$ ($r=1,\ldots,N$) and by costs per unit volume $c_r \geq 0$ ($r=1,\ldots,N$).

Here $\varepsilon = \varepsilon(u) = \frac{1}{2}(\nabla u + \nabla u^t)$ is the linear strain associated to a deformation $u: \Omega \to \mathbb{R}^3$, and each $A^{(r)}$ is a fourth-rank tensor, which must be positive and symmetric when viewed as a linear map upon second order symmetric matrices. The class of "structures" is then the set of measurable partitions of Ω into N sets, $\Omega = \Omega_1 \cup \ldots \cup \Omega_N$. To each such partition one associates an operator with piecewise constant coefficients:

$$Lu = \text{div}(\sigma(u)) = \left\{ \sum_{j=1}^{3} \frac{\partial \sigma_{ij}}{\partial x_j} \right\}_{i=1}^{3}$$

(1.3) $\quad \sigma(u) = A(x) \circ \varepsilon(u)$

$$A(x) = A^{(r)} \quad \text{if } x \in \Omega_r$$

representing the elliptic system that characterizes the elastostatics of the corresponding structure. A cost constraint has the form

(1.4) $\quad \displaystyle\sum_{r=1}^{N} c_r \text{Vol}(\Omega_r) \leq \text{constant}$.

Given a body load $f:\Omega \to \mathbb{R}^3$ and a boundary load $F:\partial\Omega \to \mathbb{R}^3$ such that $\int_\Omega f + \int_{\partial\Omega} F = 0$, the corresponding displacement is the solution u to

$$L\bar{u} + f = 0 \text{ on } \Omega, \quad \sigma(\bar{u}) \cdot \nu = F \quad \text{on } \partial\Omega,$$

where ν is the unit normal on $\partial\Omega$; the <u>compliance</u> under the load is

$$\int_{\partial\Omega} <\bar{u},F> + \int_{\Omega} <\bar{u},f> = \int_{\Omega} <\bar{\sigma},\bar{\varepsilon}>$$

where $\bar{\sigma} = \sigma(\bar{u})$, $\bar{\varepsilon} = \varepsilon(u)$ are the stress and strain of the solution.

If one were to take $A^{(0)} = 0$ in (1.2b), the problem would become one of <u>shape optimization</u>: the "zero'th material" is actually the absence of material, and one seeks not only the distribution of material but also the distribution of holes in the fixed domain Ω. The hole boundaries carry the natural boundary condition $\sigma \cdot \nu = 0$: This is the most natural choice for structural design problems. (One could, of course, impose other boundary conditions; doing so would alter the problem radically.)

The reader will have no difficulty formulating analogous classes of "structures" in the context of two-dimensional models and for other types of material behavior. In case the available materials vary continuously (instead of discretely as in 1.2b) the piecewise-constant stress strain law (1.3) must be replaced by a variable law

$$(1.3)' \qquad \sigma = A(x) \circ \varepsilon$$

where $x \mapsto A(x)$ is a measurable, tensor-valued function on Ω with values in the set of available stress-strain relations: now it is the L^∞ function $A(x)$, rather than a partition of Ω, which is the control parameter. The cost constraint (1.4) would be replaced by a condition

$$(1.4)' \qquad \int_{\Omega} G(A(x))dx \leq \text{constant}$$

where $A \mapsto G(A)$ assigns to each stress-strain law its cost per unit volume.

Although the fully 3-dimensional problems are in a sense the most fundamental ones, they are by no means the most studied. Significant work has been done, both on three-dimensional problems and on two-dimensional analogues relating to plane problems and rods in torsion, e.g. [7-12]; but the bulk of the literature addresses problems that are in various senses "reduced". For example, much effort has been devoted to optimizing the thickness distribution of beams, plates, or shells, by working directly with the corresponding fourth-order thin-structure equations. This approach has been particularly successful in one-dimensional problems - e.g. [13] - and in certain two-dimensional contexts where the limit multiplier optimization problem carries a convex structure - see [14]. In some cases, however, such analyses yield "optimal" thickness distributions with vanishing cross-sections or rapidly varying thickness. In these cases the true three-dimensional behavior is not correctly represented by the thin-structure equations.

We must add to this list of types of structures the "truss-like continua" of Michell and Prager [5] and the "grillage-like continua" of Prager and Rozvany [6]. These structures represent an idealized limit of trusses in space or stiffeners in the plane - limits, it must be said, in a sense that has never been made fully precise. The optimal layout problem in this context is among the oldest and most studied problems in the field; even here, however, much remains to be done regarding the description and effective computation of optimal structures.

The state of the art in the solution of geometry optimization problems may be summarized as follows. There are essentially three tools available: convex analysis, gradient-flow optimization, and the direct solution of first-order necessary conditions for optimality. For the most part, all these methods are restricted in their utility to integral objectives such as (1.1a,d); the sup-norm objectives (1.1b,c) have been treated successfully only in special cases (e.g. [15]).

Convex analysis is relevant to certain problems - particularly regarding limit multiplier optimization - in which the admissible materials vary continuously, and their "cost" is in some sense a convex function of their strength. A typical example is this: suppose that $\|\cdot\|$ is a norm on stresses, and one has available for each $\lambda > 0$ a material with plastic yield criterion "$\|\sigma\| = \lambda$", at cost $G(\lambda)$. Given a domain Ω and a boundary load f, the minimum cost of a structure occupying domain Ω with limit multiplier greater than or equal to one is

$$\inf_{\substack{\operatorname{div} \sigma = 0 \\ \sigma \cdot \nu = f \text{ on } \partial\Omega}} \int G(\|\sigma\|).$$

If $\lambda \to G(\lambda)$ is increasing and <u>convex</u>, this is a convex variational problem.

Most structural design optimization problems do not, however, have such a formulation - for example, a problem that deals with a discrete set of materials can not be cast in this form. (We shall see in section 3 that certain such problems may be equivalent, via relaxation, to convex ones.) In the absence of a convex structure, one may resort to the gradient - flow method, sometimes also called "optimal redesign" ([12],[16]). Given any design, a sensitivity analysis determines whether it is a stationary point in design-space for the relevant optimization problem; if not, then by following the gradient (in design space) of the objective one can iteratively improve it. Unfortunately, such methods often require a great deal of computation time, and they sometimes place implicit constraints on the geometry of the structure; on the other hand they are of great utility, since in practice it may be more important to improve a known design slightly than to find the true optimum of an abstract problem.

The third tool is the direct solution of first-order conditions. It is usually not difficult to formulate necessary conditions for a design to be optimal [3]. In the context of (1.2) - (1.4), for example, the condition that a partition be optimal for compliance under the loads f,F amounts to an extra "free boundary condition" for the displacement $\bar{u}$ at the partition boundaries $\{\partial\Omega_r\}_{r=1}^N$. There is, of course, no guarantee that this free boundary problem should have a solution. Nonetheless, the direct solution of these stationarity conditions - often called "optimality criteria" in the structures literature - has often proven fruitful. Important developments in mathematical programming have been achieved in this direction - see e.g. [17] and the "fixed-point method" described in [12].

A critical gap in all this work, however, has been the <u>absence of appropriate existence and convergence theorems</u>. While existence may be trivial for the discrete version of a given optimization problem, no convergence analysis can be expected unless one understands the behavior of the underlying continuous case; such analyses have been notably lacking. For the "generalized structures" represented by truss-like or grillage-like continua, optima apparently do exist - though even this has yet to be studied with a satisfactory level of mathematical rigor. For many other problems, however, such as the shape or material distribution problem discussed earlier, <u>there need not exist an optimal geometry</u>, at least not in the sense so far described. This realization, only gradually reached by the structural optimization community ([8,11,18, 19,41,42], see also the "nonexistence" proofs in [20]), is fundamental. In order to design computationally efficient algorithms for geometry optimization, one must first analyze the continuous problem, to determine what "generalized structures" or "effective materials" arise naturally there; then the result of <u>this</u> analysis may be approximated by discrete models, for instance using finite elements. Only by following such a procedure can rapidly oscillatory solutions - and corresponding numerical difficulties-be avoided even in the discrete problems.

§2. Homogenization and "optimal" bounds.

From the perspective of the theory of homogenization, it is clear why optimal geometries, as understood in section 1, need not exist: the problem is simply that the systems defined by (1.3), ranging over all measurable partitions of Ω, are not closed in the appropriate sense. This fact is precisely the starting point of the subject of homogenization. Given a sequence of such systems, corresponding perhaps to increasingly complicated geometries, the theory guarantees the existence of a subsequence for which the displacements converge weakly in $H^1(\Omega)$; the limit solves a new elliptic system, whose symmetries identify it as the system corresponding to an anisotropic, inhomogeneous, linearly elastic material. The subsequence can be chosen so that this convergence is uniform for loads f, F ranging in compact subsets of $H^{-1}(\Omega)$ and $H^{-1/2}(\partial\Omega)$, respectively, and so that the compliances of the solutions converge. The limiting system is called the "effective equation" or the limit of the subsequence in the sense of "homogenization" or "G-convergence". A recent review of the theory related to this sort of limiting procedure is [22]; the existence of G-limits is proven in [21]. The main point here, however, is simply this: although the original sequence arose from partitions of Ω as in (1.2,1.3), the G-limit need not correspond to any measurable partition of Ω.

Because compliances converge under passage to the G-limit, a solution to the problem of material distribution for minimal compliance will certainly exist if one extends the class of admissible structures to include the G-closure of the original class. The same is true, in fact, for any design that is preserved under G-convergence, for example the sup-norm objectives (1.1b,c). This enlargement of the class of structures is perfectly reasonable from a physical point of view: it represents the recognition that all composites which can be made from the materials on hand should be included among the admissible structures.

For problems involving shape optimization, thickness optimization of plates and shells, or elasto-plastic materials we can not be quite so precise: in these contexts the general form of the effective equations is not known. (One can, however, analyze the case of periodic local variation for some of these cases, see [23], [24], [25], [39]). The main point, however, is unaltered: in all such problems, one must allow for the possible introduction of new "effective materials" representing composites or generalized structures that can be approximated by the original class.

We have not yet demonstrated that homogenization will occur in the sort of structural optimization problem discussed in section 1; so far we have only explained why it might occur. It is not hard to construct ad hoc problems, for example minimizing $\int_\Omega |u-u_0|^2$ for suitable u_0, that display homogenization (see [20]); but proving that it occurs in a "natural" problem requires a finer analysis. This has so far been carried out primarily in two-dimensional, scalar equation model problems. The earliest, to our current knowledge, is in [8]: there one finds an analysis of the first-order optimality conditions for the distribution of material in a plane domain, so as to maximize the torsional rigidity of the corresponding rod, under various cost contraints. For certain such problems this procedure leads to a "contradiction": the reason, explained only somewhat vaguely in [8], is that for those formulations the optimal configuration is a homogenized one. The recent paper [11] reexamines the problem, extending the class of admissible structures to laminar composites (still not the full G-closure), and shows that the new first-order optimality conditions so obtained can indeed be satisfied.

Recent work by the authors ([26],[29]; see also section 3 below) deals with some model problems involving the distribution of holes or of several materials in a plane domain, to optimize the torsional rigidity, compliance, or limit multiplier of the corresponding rod under given loading. For these problems one finds that homogenization almost always occurs; the optimal structures are a sort of fibered composite.

Homogenization has also been detected in problems involving the thickness distribution of an elastic plate. In numerical work [18], and in the associated analyses [19] and [11], Cheng, Olhoff, and Lurie, <u>et al</u>. have shown that for certain compliance optimization problems under transverse loads, with constraints on the maximum and minimum thickness and on the total weight, an "optimal" structure must be a generalized one.

There are, it should be noted, apparently similar problems where homogenization does not occur: for example, the holes constructed in [28] represent, according to the results of [15], configurations that minimize the stress concentrations in certain problems of plane stress. One does not yet know how to predict, for a particular problem, whether or not homogenization will occur. There is, however, one characteristic shared by many of the problems now known to exhibit it: in them, the "unknown boundaries" carry Neumann or natural transmission boundary conditions. Another class of geometry problems is often studied in the context of heat conduction, electrostatic capacity, and cavitation: these problems place a Dirichlet condition on the "unknown boundary". Existence theorems (without homogenization) for some problems of this type may be found in [29], [30], and the references therein.

We can now make explicit the relationship between "optimal bounds" for effective materials and the optimization of structural geometry.

To find optimal configurations in a problem where homogenization can occur, one needs to know

        i) The G-closure of the original class of equations (without cost constraints)

        ii) The minimum cost per unit volume of each element of this G-closure.

It certainly suffices for this to know sharp bounds (traditionally called "optimal bounds") for the relevant effective equations, as a function of relative volume fractions. Such bounds would specify, for given $\{\rho_j\}_{i=1}^{N}$ ($0 \leq \rho_j \leq 1$, $\sum \rho_j = 1$), which anisotropic, homogeneous, elastic effective materials can be obtained using volume fraction ρ_j of the jth component.

In special cases one might be able to use an invariance in the problem to reduce the information required. For example, according to the analysis in [31], one could maximize the torsional rigidity of a rod by knowing the minimum cost per unit volume, as a function of the <u>largest eigenvalue</u> of the "stress-strain law" (which in this context is a symmetric 2×2 matrix).

The idea that one can create new effective materials by mixing given ones on a microscopic scale is certainly not new. The problem of determining optimal bounds, too, is an old one: major contributions have been made in recent years by Hashin, Shtrikman, Bergman, Beran, Hill, and Willis, to name but a few. For the study of geometry optimization problems, however, it is very important that no geometric constraints - such as periodicity or randomness - be placed upon the admissible configurations; and it is crucial that the bounds be attainable.

The best tool for proving fully rigorous bounds without geometric hypotheses seems to be the machinery of compensated compactness introduced by Murat and Tartar [33,34,35]. Recent work by Tartar (presented at this conference but not yet published) includes fully rigorous proofs of such bounds, and proofs of their attainability, in the case of the scalar Laplace equation, for mixtures of two isotropic dielectrics. Further results on more complex problems are sure to come.

Much remains to be done, however, to bridge the gap between homogenization theory and structural optimization practice. The following list of open problems represents a summary of part of the job that remains

        1) Find sharp bounds for effective equations in the context of plane strain and three dimensional elastostatics.

        2) Explore the effect on the optimization problem of replacing the correct bounds on effective equations by approximate ones (which might, for example, be produced

by an algorithm for solving question 1).
3) Determine the effective equations for composites that include holes or elasto-plastic inclusions; also, for the variational inequality problems that arise from the optimal design of bearings, see e.g. [36].
4) What regularity can one expect for the coefficients in the effective equation that represents an optimal configuration?
5) Design an efficient finite element scheme for using sharp effective-equation bounds to compute locally optimal configurations.
6) Develop a mathematically rigorous theory of Michell's "truss-like continua" and Prager-Rozvany's "grillage-like continua," including a proof of the existence and regularity of optimal geometries and a description of their behavior.

§3. Structural Optimization and Relaxation.

We have recently introduced another method for answering certain questions of geometry optimization. Our approach, which involves the relaxation of an appropriate nonconvex variational problem, is less general than the one sketched in section 2; when it applies, however, it gives more explicit information about the optimal configurations.

In principle the approach applies to a variety of elasticity problems, involving either shape optimization or optimal material distribution, in the presence of compliance constraints under each of a finite number of loads. It applies equally well to a plasticity problem in which the constraints are on the limit multipliers.

In practice, however, we have been able to carry it out only for a much more restricted class of problems: those modeled by a scalar equation, in the presence of a single load. Even these simple cases lead to interesting results; a full discussion is currently in preparation, and examples may be found in [26,27]. Here we present a brief and nontechnical exposition of the method in a special case, the shape optimization of a rod in antiplane shear. With that discussion as motivation, we then describe some open problems whose resolution would allow the extension of the method to a more practical context.

Let Ω be a bounded domain in $\mathbb{R}^2$, and $f: \partial\Omega \to \mathbb{R}$ with $\int_{\partial\Omega} f = 0$. We imagine that Ω is the cross-section of an infinite elastic rod, with

load f directed lengthwise along its boundary, uniformly along the
length of the rod. (See figure 1.) After some reductions, and ignoring

Figure 1.

elastic constants, the compliance of the rod per unit length can be
expressed as

$$c(\Omega) = \inf_{\sigma \in X(f)} \int_\Omega |\sigma|^2 ,$$

where

$$X(f) = \{\sigma \in L^2(\Omega; \mathbb{R}^2): \quad \text{div } \sigma = 0 \text{ on } \Omega, \ \sigma \cdot \nu = f \text{ on } \partial\Omega.\}.$$

We assume that Ω, f are regular enough that $X(f)$ be nonempty.

The shape-optimization problem is this: how should one remove
material from the interior of Ω so as to increase the compliance (i.e.
weaken the structure) as little as possible? Whenever a hole (a closed
set) $H \subset \Omega$ is removed, the newly formed boundary ∂H is to be unloaded,
so the resulting structure has compliance

$$c(\Omega \sim H) = \inf_{\substack{\sigma \in X(f) \\ \sigma = 0 \text{ a.e. on } H}} \int_\Omega |\sigma|^2 .$$

The constraint "$\sigma = 0$ a.e. on H" serves to impose the condition $\sigma \cdot \nu = 0$
on ∂H. Introducing a parameter $\alpha > 0$, we consider the modified problem

(3.1) $\quad \inf_{H \subset \Omega} \text{Area }(H) + \alpha \ c(\Omega \sim H) ,$

where H ranges over closed subsets of Ω. If there is an extremal set
H_0 for (3.1), then it certainly must minimize $c(\Omega \sim H)$ for fixed Area(H);

This would be an optimal geometry in the conventional sense. If there is no extremal, a minimizing sequence for (3.1) must in any case consist of sets that nearly minimize $c(\Omega \sim H)$ for their areas, and we can expect to find an optimal "generalized structure." (One can also treat α formally as a Lagrange multiplier, in order to find minimum - area geometries with fixed compliance; this more technical aspect will be presented elsewhere.)

Ignoring the fact that $\{x:\sigma(x)=0\}$ need not be closed for arbitrary $\sigma \in X(f)$ (this technical point, too, will be addressed elsewhere), one finds that (3.1) is essentially the same as

$$(3.2) \qquad \inf_{\sigma \in X(f)} \int_\Omega (1_{\{\sigma \neq 0\}} + \alpha |\sigma|^2) ,$$

where 1_Y denotes the characteristic function of a set $Y \subset \Omega$. This, then, is the starting point of our method: the shape optimization problem (3.1) can be reformulated as a variational problem over $X(f)$ with the discontinuous integrand

$$G_\alpha(\sigma) = \begin{cases} 1 + \alpha|\sigma|^2 & \text{if } \sigma \neq 0 \\ 0 & \text{if } \sigma = 0 \end{cases}$$

If in (3.2) σ ranged over all L^2 vector fields then the infimum would be unchanged by replacing G_α with its convex hull

$$F_\alpha(\sigma) = \begin{cases} 1 + \alpha|\sigma|^2 & |\sigma| \geq \alpha^{-1/2} \\ 2\alpha^{1/2}|\sigma| & |\sigma| \leq \alpha^{-1/2} \end{cases}$$

(see figure 2). Being convex, $\int F_\alpha$ is weakly lower semicontinuous, so it has a minimizer in $X(f)$; given any minimizer for $\int F_\alpha$, one can construct a minimizing sequence for $\int G_\alpha$ by introducing appropriate small-scale oscillations in σ: this passage from $\int G_\alpha$ to its lower semicontinuous hull $\int F_\alpha$ is called relaxation (see e.g. [38]). In the present context more care is required: we must respect the constraints $\sigma \cdot \nu = f$ and div $\sigma = 0$. However, using methods related to those of [38, chapter X] one may show that the constraints do not affect the relaxation. Hence (3.2) is equivalent to the relaxed problem

$$(3.3) \qquad \inf_{\sigma \in X(f)} \int_\Omega F_\alpha(\sigma) .$$

A more careful examination of the relaxation procedure shows that to any extremal σ_0 of (3.3) one can associate a minimizing sequence

$\{\sigma_i\}_{i=1}^{\infty}$ for (3.2) by choosing - roughly speaking - σ_i parallel to σ_0, with $\sigma_i = \sigma_0$ where $|\sigma_0| \geq \alpha^{-1/2}$ or $|\sigma_0| = 0$, but $|\sigma_i|$ oscillating rapidly between 0 and $\alpha^{-1/2}$ where $0 < |\sigma_0| < \alpha^{-1/2}$; the sets where $\sigma_i = 0$ constitutes a minimizing sequence for (3.1). It is not hard to construct examples where $0 < |\sigma_0| < \alpha^{-1/2}$ on a set of positive measure; for these problems an optimal geometry exists only in a generalized sense - and we see that the appropriate class of generalized structures is a sort of "fibred continuum."

The analogous plasticity problem is the optimal removal of area from Ω given a constraint on the <u>limit multiplier</u> of the load f. Making appropriate reductions and again ignoring a material constant, the limit multiplier of the load upon $\Omega \sim H$ is

$$L(\Omega \sim H) = \sup\{t: \exists \sigma \in X(f), \sigma = 0 \text{ a.e. on } H, |\sigma| \leq t^{-1}\}.$$

Therefore the geometry problem

(3.4) $\quad\quad\quad \inf_{\substack{H \subset \Omega \\ L(\Omega \sim H) \geq \lambda}} \text{Area } (\Omega \sim H)$

reduces to the variational problem

(3.5) $\quad\quad\quad \inf_{\substack{\sigma \in X_f \\ |\sigma| \leq \lambda^{-1}}} \int_\Omega 1_{\{\sigma \neq 0\}}$

Once again, the shape optimization problem can be solved by relaxing (3.5); this time the relaxed problem is

(3.6) $\quad\quad\quad \inf_{\substack{\sigma \in X_f \\ |\sigma| \leq \lambda^{-1}}} \lambda \int_\Omega |\sigma|$,

and the appropriate class of generalized structures is, once again, the "fibred continua". (See figure 3).

Figure 2.

Figure 3.

One can say quite a bit more about the relaxed problems (3.3) and (3.6), using convex duality and other tools; proofs and further analysis, as well as a treatment of other model problems by a similar method, will be presented elsewhere. Here we focus instead on how to extend the method to a wider class of problems.

Our approach does seem to require that the constraints be closely related to the variational structure of the analysis problem; for the moment, it seems restricted (unlike the approach of section 2) to compliance problems in elasticity and to limit analysis problems in plasticity. There is, however, no difficulty in formulating analogues of (3.2) or (3.5) in a fully three-dimensional context, and in the presence of any finite number of loads. The obstacle to solving such problems lies in the computation of the correct relaxed problem. (We are convinced that it will not, in most cases, be the convex hull of the discontinuous integrand!) Here is a list of four specific versions of this problem.

1) (Antiplane shear of an elastic rod, as above, but under several different loads.) Let $f_1, \ldots, f_N : \partial\Omega \to \mathbb{R}$ with $\int_{\partial\Omega} f_i = 0$ for each i, and fix $\{\alpha_i\}_{i=1}^N$, $\alpha_i \geq 0$. Find the relaxation of

$$\inf_{\sigma_i \in X(f_i)} \int_\Omega \chi(\sigma_1, \ldots, \sigma_N) + \sum_{i=1}^N \alpha_i |\sigma_i|^2$$

where

$$\chi(\sigma_1, \ldots, \sigma_N) = \begin{cases} 1 & \text{if } \sigma_i \neq 0 \text{ for some } i \\ 0 & \text{if } \sigma_1 = \ldots = \sigma_N = 0. \end{cases}$$

2) (The same problem in a limit analysis formulation): for $\{f_i\}_{i=1}^N$ as above, find the relaxation of

$$\inf_{\substack{\sigma_i \in X(f_i) \\ |\sigma_i| \leq 1}} \int_\Omega \chi(\sigma_1, \ldots, \sigma_N)$$

3) (Three-dimensional shape optimization, minimizing a linear combination of volume and compliance.) Let E denote the space of 3×3 symmetric tensors, and let $A : E \to E$ be an elastic stress-strain law, with $\|\sigma\|^2 = \langle \sigma, A^{-1}\sigma \rangle$ the corresponding "strain energy" norm on stresses. Let

$\Omega \subset \mathbb{R}^3$, $f:\partial\Omega \to \mathbb{R}^3$, and assume $\int_{\partial\Omega} f = 0$. For $\alpha > 0$, find the relaxation of

$$\inf_{\sigma \in X'(f)} \int_\Omega 1_{\{\sigma \neq 0\}} + \alpha \|\sigma\|^2 ,$$

where $X'(f) = \{\sigma \in L^2(\Omega;E): \text{div } \sigma = 0 \text{ on } \Omega, \sigma \cdot \nu = f \text{ on } \partial\Omega\}$.

4) (The analogue of 3, for plastic limit analysis.) Let K be a convex subset of E containing a neighborhood of the origin. Find the relaxation of

$$\inf_{\substack{\sigma \in X'(f) \\ \sigma(x) \in K \text{ a.e.}}} \int_\Omega 1_{\{\sigma \neq 0\}} ,$$

where Ω, f and $X'(f)$ are as in (3).

References

1) Lagrange, J.L., "Sur la figure des colonnes", <u>Miscellanea Taurinensia</u> <u>V</u>, 1770-1773, pg. 123.
2) Michell, A.G.M., "The limits of economy of material in frame structures", <u>Phil. Mag.</u> S6, Vol 8, No.47, pp 589-597.
3) Haug, E.J., and Cea, J., Proceedings of NATO ASI on optimization of distributed parameter structures, Iowa City, 1980. Sithoff and Nordhoff, to appear.
4) Sawzuk, A. and Mroz, Z., <u>Optimization in structural design</u> (Proceedings of 1973 IUTAM symposium in Warsaw, Poland), Springer-Verlag, 1975.
5) Prager, W., <u>Introduction to structural optimization</u>, International Centre for Mechanical Science, Udine, Courses and Lectures no. 212, Springer-Verlag, 1974.
6) Rozvany, G.I.N., <u>Optimal design of flexural systems</u>, Pergamon Press, 1976.
7) Cea, J. and Malanowski, K., "An example of a max-min problem in partial differential equations", <u>SIAM J. Control</u> vol 8, 1970, pp. 305-316.
8) Klosowicz, B. and Lurie, K.A., "On the optimal nonhomogeneity of a torsional elastic bar", <u>Arch. of Mechanics</u>, vol 24, 1971, pp 239-249.
9) Mroz, Z., "Limit analysis of plastic structures subject to boundary variations", <u>Arch. Mech. Stos.</u> vol 15, 1963, pp. 63-75.
10) Zavelani, A., Maier, G., and Binda, L., "Shape optimization of plastic structures by zero-one programming", in IUTAM Warsaw Symposium, 1973, see reference 3.
11) Lurie, K.A., Fedorov, A.V., and Cherkaev, A.V., "Regularization of optimal design problems for bars and plates and elimination of con-

tradictions within the necessary conditions of optimality", <u>Journal of Opt. Th. and Appl.</u>, to appear, 1982

12) Cea, J., "Shape optimal design: problems and numerical methods", in Proc. of NATO-ASI on Optimization of Distributed Parameter Structures, Iowa City, 1980

13) Olhoff, N., "Optimization of columns against buckling" in Proceedings of NATO-ASI on optimization of distributed parameter structures, 1980.

14) Cinquini, C. and Mercier, B., "Minimal cost and elastoplastic structures", <u>Meccanica</u> vol 11, no 4, 1976, pp 219-226.

15) Wheeler, L., "On the role of constant - stress surfaces in the problem of minimizing elastic stress concentration.", <u>International J. of Solids and Structures</u> vol 12, 1967, pp 779-789.

16) Olhoff, N. and Taylor, J.E., "On optimal structural remodeling", <u>J. Opt. Theory and Applic.</u> Vol 27, 1979, pp 571-582.

17) Fleury, C. and Schmit, L.A., "Primal and dual methods in structural optimization," <u>J. Structural Div. ASCE</u> vol 106, 1980, pp 1117-1133.

18) Cheng, K.T. and Olhoff, N., "An investigation concerning the optimal design of solid elastic plates", <u>Int. J. of Solids and Structures</u>, to appear.

19) Cheng, K.T. and Olhoff, N., "Regularized formulation for optimal design of axisymmetric plates", to appear.

20) Murat, F., "Contre-examples pour divers problemes ou le control intervient dans les coefficients", <u>Annali di Mat. Pura ed Appl.</u> Ser 4, vol 112-113, 1977.

21) Simon, L., "On G-convergence of elliptic operators," Indiana Univ. Math. Journal vol 28, pp 587-594.

22) De Giorgi, E., "Convergence problems for functionals and operators", in Proc. of the International Meeting on Recent Methods in Nonlinear Analysis, Rome, 1978; De Giorgi, Magenes, & Mosco editors, Pitagora Editrice, Bologna, 1980.

23) Duvaut, G., "Comportement microscopique d'une plaque perforée périodiquement", to appear.

24) Cioranescu, D. and Saint Jean Paulin, J., "Homogenization dans des ouverts à cavités," <u>C.R. Acad. Sci. Paris</u> A, vol 284 (1977) pp 857-860.

25) Carbone, L., "Sur un problème d'homogénéisation avec des constraints sur le gradient", <u>J. Math. Pures et Appl.</u> 58, 1979, pp 275-297.

26) Strang, G. and Kohn, R., "Optimal design of cylinders in shear", to appear in proceedings of the 1981 MAFELAP Conference, Brunel University.

27) Kohn, R. and Strang, G., "Optimal design for torsional rigidity", to appear in proceedings of the conference on Mixed and Hybrid Methods in Finite Element Methods, Atlanta, 1981.

28) Cherepanov, G.P., "Inverse problems of the plane theory of elasticity",

P.M.M. vol 38, no 6, 1974, pp 963-979.

29) Acker, A., "Interior free boundary problems for the Laplace equation", Arch. Rat. Mech. Anal. vol 75, 1981, pp 157-168.

30) Alt, H.W. and Caffarelli, L.A., "Existence and regularity for a minimum problem with free boundary", J. Riene Angew. Math. 325 (1981) pg 105.

31) Banichuk, N., Doklady Akad. Nauk USSR vol 242, pp 1042-1045, 1978.

32) Jouron, C., "Sur un problème d'optimisation où la constrainte porte sur la fréquence fondamentale", RAIRO Analyse Numerique vol 12, 1978, pp 349-376.

33) Murat, F., "Compacité par compensation", Ann. Scuola Norm. Sup Pisa vol 5, 1978, pp. 489-507.

34) Murat, F., "Compacite par compensation II", in proc. of the International Meeting on Recent Methods in Nonlinear Analysis, De Giorgi, Magenes, Mosco editors, Pitagora Editrice, Bologna, 1980.

35) Tartar, L., "Estimation de coefficients homogeneises", Springer Lect. Notes in Math. vol 704, pp 364-373.

36) Benedict, R.L., "Optimal design for elastic bodies in contact", in proceedings of the NATO-ASI on optimization of distributed parameter structures, (ref.3).

37) Banichuk, N.V.; Kartvelishvili, V.M.; and Mironov, A.A., "Optimization problems with local performance criteria in the theory of plate bending", Mechanics of Solids, 1978, no 1.

38) Ekeland, I. and Temam, R., Convex Analysis and Variational Problems, North-Holland, 1976.

39) Vogelius, M.; Kohn, R.; Papanicolaou, G., "Effective equations for plates and beams with rapidly varying thickness", to appear.

40) Olhoff, N.; Lurie, K.A.; Cherkaev, A.V.; Fedorov, A.V.; "Sliding Regimes and Anisotropy in Optimal Design of Vibrating Axisymmetric Plates", Int. J. Solids Structures, to appear.

41) Raitum, U.E., "On optimal control problems for linear elliptic equations", Soviet Math. Dokl. Vol 20, pp 129-132, 1979.

42) Raitum, U.E., "The extension of extremal problems connected with a linear elliptic equation", Soviet Math. Dokl. Vol 19, pp 1342-1345, 1978.

Coherent Medium Approach to Hopping Conduction

M. Lax
*Department of Physics, The City College of the City
University of New York, New York, N.Y. 10031
and Bell Laboratories, Murray Hill, N.J. 07974*

T. Odagaki
*Department of Physics, The City College of the City
University of New York, New York, N.Y. 10031*

Contents

1. Introduction
2. Exact coherent medium I. - Multiple scattering formalism
3. Exact coherent medium II. - Projection operator method and the CTRW
4. A simple example of the exact coherent medium
5. The coherent medium approximation
6. Other approximations
7. Bond-percolation model in a chain I. - Exact solution
8. The coherent medium approximation in one-dimensional chains
9. Binary jump rate in a chain
10. Bond-percolation model in a chain II. - Approximate treatment
11. Chains with positionally-random localized centers
12. Bond-percolation model in a simple cubic lattice
13. Comparison with a computer simulation - Miller-Abrahams-jump rate
14. Impurity conduction
15. Summary

1. Introduction

Stochastic transport in disordered or amorphous materials consists of hopping of carriers from one localized center to another with the assistance of other degrees of freedom such as phonons. An ac conductivity due to the hopping process was first observed in doped silicon by Pollak and Geballe[1] in 1961. The typical and a striking feature of the ac conductivity is that the observed data at various temperatures and frequencies for a given doping level can be fitted to a single curve by a relevant scaling of both the conductivity and the frequency, and the so-called ac part (Real$\{\sigma(\omega)-\sigma(0)\}$) of the conductivity shows a power law dependence on the frequency at low frequencies. [See later Figs. 14.1 and 14.2.] The electric transport of this system is believed to be caused by electron hopping from a neutral donor to a donor ionized by a compensating acceptor. [See Fig. 1.1.]

A similar mechanism of conduction has been employed to explain low-frequency ac conductivity in many other systems such as doped VO_2,[2] Na-βAl_2O_3,[3] MoO_3,[4] Te-doped Se,[5] and spinel-type MnCoNiCu complex oxides[6], amorphous chalcogenides[5,7] and so on. In each case, carriers (electrons or ions) are assumed to move from one localized center to another by a hopping process. Environmental fluctuations and the assistance of other degrees of freedom makes the site-to-site motion of carriers stochastic.

Now, let us introduce a conditional probability $P(\mathbf{s},t|\mathbf{s}_0,0)$ of finding a moving carrier at a point $\mathbf{s}$ at time t if it was at a point $\mathbf{s}_0$ at $t=0$. It is natural to assume that $P(\mathbf{s},t|\mathbf{s}_0,0)$ obeys a random walk equation

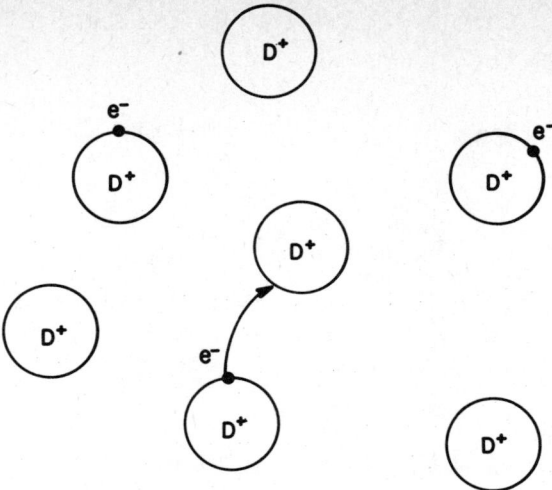

Fig. 1.1. Schematic illustration of hopping motion of an electron in a doped semiconductor. An electron jumps from a neutral donor to a donor ionized by a compensating acceptor.

$$\frac{\partial P(\mathbf{s},t)}{\partial t} = -\Gamma_\mathbf{s} P(\mathbf{s},t) + \sum_{\mathbf{s}'\neq\mathbf{s}} w_{\mathbf{s}\mathbf{s}'} P(\mathbf{s}',t) \; , \qquad (1.1)$$

where the total decay rate out of $\mathbf{s}$ is given by

$$\Gamma_\mathbf{s} = \sum_{\mathbf{s}'\neq\mathbf{s}} w_{\mathbf{s}'\mathbf{s}} \qquad (1.2)$$

and for simplicity of notation we have omitted listing the initial conditions. The definition of $\Gamma_\mathbf{s}$ Eq. (1.2) insures conservation of probability $\sum_\mathbf{s} P(\mathbf{s},t|\mathbf{s}_0,0)=1$. We do not include any traps in the present paper. The jump rate $w_{\mathbf{s}'\mathbf{s}}$ from site $\mathbf{s}$ to $\mathbf{s}'$ is assumed to have an appropriate dependence on the distance $|\mathbf{s}'-\mathbf{s}|$ between two sites. As a simplified or tractable model for positionally random systems, hopping conduction on lattices will be studied, where the hopping sites $\{\mathbf{s}\}$ form a regular array of lattice points and a suitable distribution of jump rates is introduced to replace randomness in jump rates induced by positional disorder. In a sense, we study a random walk on a lattice in a random environment.

For later convenience, we introduce the Laplace transform of the probability $P(\mathbf{s},t|\mathbf{s}_0,0)$

$$\tilde{P}(\mathbf{s},u|\mathbf{s}_0) = \int_0^\infty e^{-ut} P(\mathbf{s},t|\mathbf{s}_0,0) \, dt \qquad (1.3)$$

which obeys

$$(u+\Gamma_\mathbf{s})\tilde{P}(\mathbf{s},u|\mathbf{s}_0) - \sum_{\mathbf{s}'\neq\mathbf{s}} w_{\mathbf{s}\mathbf{s}'} \tilde{P}(\mathbf{s}',u|\mathbf{s}_0) = \delta(\mathbf{s},\mathbf{s}_0) \; , \qquad (1.4)$$

$\delta(\mathbf{s},\mathbf{s}_0)$ being a Kronecker δ-function.

A formal solution of Eq. (1.4) is readily given by a matrix element

$$\tilde{P}(\mathbf{s},u|\mathbf{s}_0) = \{(u\hat{1}-\hat{H})^{-1}\}_{\mathbf{s},\mathbf{s}_0} , \qquad (1.5)$$

where a matrix $\hat{H}$ is defined by

$$\hat{H}_{\mathbf{s},\mathbf{s}'} = w_{\mathbf{s}\mathbf{s}'} \quad \text{for } \mathbf{s} \neq \mathbf{s}' \qquad (1.6)$$

and

$$\hat{H}_{\mathbf{s},\mathbf{s}} = -\Gamma_{\mathbf{s}} . \qquad (1.7)$$

Equation (1.5) makes $\tilde{P}(\mathbf{s},u|\mathbf{s}_0)$ behave as the matrix element of a propagator and $\hat{H}$ plays the role of the corresponding Hamiltonian matrix.[8] Thus, we may call $(u\hat{1}-\hat{H})^{-1}$ a random walk propagator.

We are primarily interested in the ac conductivity due to the hopping motion of carriers described by Eq. (1.1). It has been shown[9,10] that when $kT \ll \hbar\omega$ the ac conductivity is reducible to the generalized Einstein relation

$$\sigma(\omega) = \frac{ne^2}{kT} D(\omega) , \qquad (1.8)$$

and the frequency-dependent diffusion constant is written in terms of the second spatial moment of $\tilde{P}(\mathbf{s},i\omega|\mathbf{s}_0)$ as

$$D(\omega) = -\frac{\omega^2}{2d} \sum_{\mathbf{s}} < (\mathbf{s}-\mathbf{s}_0)^2 \tilde{P}(\mathbf{s},i\omega|\mathbf{s}_0) > . \qquad (1.9)$$

Here, d is the dimensionality of the system, n denotes the number density of carriers with charge e, k is the Boltzmann constant, T is the absolute temperature and $<\cdots>$ denotes the average over the ensemble of possible $w_{\mathbf{s}\mathbf{s}'}$. Note that the average over the initial site $\mathbf{s}_0$ has been omitted in Eq. (1.9), because the ensemble average eliminates the dependence of the summand on $\mathbf{s}_0$. If the sites $\{\mathbf{s}\}$ form a regular lattice, $(\mathbf{s}-\mathbf{s}_0)^2$ can be taken outside of the ensemble average. (The randomness resides in the jump rates.) Eventually, in order to obtain the conductivity we must evaluate the ensemble average of Eq. (1.5). The dielectric constant $\epsilon(\omega)$ at low frequencies is expressible in terms of $\sigma(\omega)$.

Another interesting quantity is the decay of the carrier from its initial site which is simply described by $P(\mathbf{s}_0,t|\mathbf{s}_0,0)$ itself. Although there is extensive literature[11] on spectral diffusion and excitation transfer which are governed by the time decay function $P(\mathbf{s}_0,t|\mathbf{s}_0,0)$, we will not discuss $P(\mathbf{s}_0,t|\mathbf{s}_0,0)$ except for a few cases.

In passing, we define the dimensionless conductivity or diffusion constant by

$$\tilde{\sigma}(\omega) = \tilde{D}(\omega) \equiv \sigma(\omega)/(ne^2a^2w_0/kT) , \qquad (1.10)$$

using scaling parameters a and w_0 of distance and frequency, respectively, which will appear in individual problems.

The present paper is organized as follows. In Sections 2 and 3 we present two exact expressions for the coherent medium which is defined through an average of the random walk propagator; one is the multiple scattering description by Lax[12,13] and another is a formal proof of the validity of the continuous time random walk (CTRW) description of Scher and Lax.[9] This was first given in t space by Klafter and Silbey,[14] using projection techniques developed by Zwanzig.[15] A shorter proof will be presented here in the Laplace transformed space. The multiple scattering method provides the starting point of the subsequent coherent medium approximation (CMA). The exact coherent medium is easily obtained at the high frequency limit, which is given in Section 4. In Section 5, we explain a generalized application of the coherent potential approximation[12,16] to obtain the averaged random walk propagator. In Section 6, we discuss other approximations proposed to treat the random walk equation with random jump rate. Sections 7-11 are devoted to the hopping conduction on one-dimensional chains. First, in Section 7 a bond-percolation model where a jump rate has a finite probability of vanishing is solved exactly to provide a standard against which approximate methods can be tested. Then, general results obtained by the coherent medium approximation are given in Section 8. The distribution of jump rate is classified into six categories according to the first and second moments of the inverse of the jump rate which determine the behavior of the ac conductivity in the vicinity of the static limit. An example is given for a case in which the dc conductivity vanishes even though a carrier can diffuse away infinitely from its initial position. The ac conductivity is evaluated in the present approximation for three types of the distribution of the jump rate, (1) binary jump rate [Section 9], (2) bond-percolation model [Section 10], and (3) a distribution which is derived to simulate a positionally random distribution of hopping sites [Section 11]. Applications to the three dimensional system are given in Sections 12-14. Section 12 treats the bond-percolation model and several critical behaviors at the percolation threshold are discussed. A comparison with computer simulation is given in Section 13. The method is also applied to reproduce the low-frequency conductivity in doped semiconductors (Section 14). A short summary is given in Section 15.

2. Exact coherent medium I - Multiple scattering formalism

An elegant multiple scattering formalism for wave propagation in random media was first given by Lax.[12,13] The idea can be employed to express the inverse matrix in Eq. (1.5) by an integral equation, even though there is no physical wave at all.

Let us define a matrix $\hat{G}$ (the random walk propagator) by

$$\hat{G} \equiv (u\hat{1} - \hat{H})^{-1} . \qquad (2.1)$$

For any matrix $\hat{H}_1$, $\hat{G}$ can be expressed as

$$\hat{G} = \hat{G}_1 + \hat{G}_1 \hat{V} \hat{G} , \qquad (2.2)$$

where

$$\hat{G}_1 = (u\hat{1} - \hat{H}_1)^{-1} , \qquad (2.3)$$

and

$$\hat{V} = \hat{H} - \hat{H}_1 . \qquad (2.4)$$

Now, suppose the difference matrix $\hat{V}$ can be written as a sum of submatrices.

$$\hat{V} = \sum_i \hat{V}_i . \qquad (2.5)$$

Here, i stands for an index which represents the unit. For $\hat{H}$ defined by Eqs. (1.6) and (1.7), for example, the unit consists of a large matrix whose nonvanishing elements are those connected to a pair of sites. Then, Eq. (2.2) can be converted into the following equations of multiple scattering form:

$$\hat{G} = \hat{G}_1 + \sum_j \hat{G}_1 \hat{T}_j \hat{G}^j , \qquad (2.6)$$

$$\hat{G}^j = \hat{G}_1 + \sum_{i \neq j} \hat{G}_1 \hat{T}_i \hat{G}^i , \qquad (2.7)$$

where the transition matrix obeys

$$\hat{T}_i = \hat{V}_i + \hat{V}_i \hat{G}_1 \hat{T}_i . \qquad (2.8)$$

The proof of these equations is found in Refs. 12 and 13. So far the choice of $\hat{H}_1$ is completely arbitrary and no approximation has been made. The best choice of $\hat{H}_1$ will be such that $\hat{H}_1$ describes the averaged properties of the system. The ensemble average of Eq. (2.6) leads to an exact definition of a coherent medium $\hat{\Sigma}$ through

$$<\hat{G}> = \hat{G}_c \equiv (u\hat{1} - \hat{\Sigma})^{-1} , \qquad (2.9)$$

where $\hat{\Sigma}$ is chosen to satisfy

$$\sum_j <\hat{T}_j \hat{G}^j> = 0 . \qquad (2.10)$$

Here, the matrix $\hat{H}_1$ in Eqs. (2.6) and (2.7) is replaced by the as yet unknown coherent matrix $\hat{\Sigma}$.

3. Exact coherent medium II - Projection operator method and the CTRW

Recently Klafter and Silbey[14] formally proved that the original random walk problem in a random environment can be rigorously reduced to a CTRW calculation of the Scher-Lax type. We shall supply a simpler version of the Klafter-Silbey derivation. Like their proof, the derivation is based on projection operator techniques. The simplicity is achieved by performing the derivation completely in the Laplace transform domain rather than working in the time domain and then transforming to the Laplace domain.

We introduce a projection operator **P** in the manner so heavily exploited by Zwanzig:[15]

$$\mathbf{P}(\text{anything}) = <\text{anything}> = \text{average of anything} . \qquad (3.1)$$

Equation (2.1) can be written in the form

$$(u\hat{1} - \hat{H}) \hat{G} = \hat{1} . \qquad (3.2)$$

We are trying to evaluate $<\hat{G}> = \mathbf{P}\hat{G}$. If we also introduce the complementary projection operator

$$\mathbf{Q} = 1 - \mathbf{P} , \qquad (3.3)$$

$\mathbf{P}\hat{G}$ can be obtained using techniques previously introduced to deal with degenerate perturbation theory.[17] Multiplication of Eq. (3.2) by **P** yields

$$u\mathbf{P}\hat{G} - \mathbf{P}\hat{H}\hat{G} = \hat{1} \qquad (3.4)$$

or

$$(u\hat{1} - \mathbf{P}\hat{H}\mathbf{P})\mathbf{P}\hat{G} = \hat{1} + \mathbf{P}\hat{H}\mathbf{Q}\hat{G} , \qquad (3.5)$$

where the latter form was obtained by replacing $\hat{H}$ by $\hat{H}(\mathbf{P}+\mathbf{Q})$ and an extra **P** has been introduced at no extra charge since $\mathbf{P}^2 = \mathbf{P}$. Since we want to find $\mathbf{P}\hat{G} = <\hat{G}>$, we must eliminate $\mathbf{Q}\hat{G}$. To obtain $\mathbf{Q}\hat{G}$, multiply Eq. (3.2) by **Q** and use $\mathbf{Q}\hat{1} = (1-\mathbf{P})\hat{1} = \hat{1}-\hat{1} = 0$ to obtain

$$(u\hat{1} - \mathbf{Q}\hat{H}\mathbf{Q})\mathbf{Q}\hat{G} = \mathbf{Q}\hat{H}\mathbf{P}\hat{G} . \qquad (3.6)$$

Equation (3.6) may now be solved for $\mathbf{Q}\hat{G}$ and the result be inserted into Eq. (3.5) to obtain

$$[u\hat{1} - \hat{\Sigma}(u)] \mathbf{P}\hat{G} = \hat{1} , \qquad (3.7)$$

where the "coherent medium" $\hat{\Sigma}(u)$ is given by

$$\hat{\Sigma}(u) = \mathbf{P}\hat{H}\mathbf{P} + \mathbf{P}\hat{H}\mathbf{Q}(u\hat{1}-\mathbf{Q}\hat{H}\mathbf{Q})^{-1}\mathbf{Q}\hat{H}\mathbf{P} , \qquad (3.8)$$

and an extra **Q** has been inserted. Also, since $\mathbf{PQ} = 0$ we can write

$$\mathbf{P}<\hat{H}>\mathbf{Q}\hat{A}\mathbf{P} = 0 , \qquad (3.9)$$

where $\hat{A}$ is any matrix. Thus we can write

$$P\hat{H}Q\hat{A}P = P\delta\hat{H}Q\hat{A}P = P\delta\hat{H}Q\delta\hat{A}P , \tag{3.10}$$

where

$$\delta\hat{H} \equiv \hat{H} - <\hat{H}> ; \delta\hat{A} \equiv \hat{A} - <\hat{A}> . \tag{3.11}$$

Equation (3.9) can therefore be written in a notation similar to that used by Klafter and Silbey

$$\hat{\Sigma}(u) = P\hat{H}P + P\delta\hat{H}(u\hat{1}-Q\hat{H}Q)^{-1}\delta\hat{H}P , \tag{3.12}$$

or

$$\hat{\Sigma}(u) = <\hat{H}> + <\delta\hat{H}(u\hat{1}-Q\hat{H}Q)^{-1}\delta\hat{H}> . \tag{3.13}$$

The principal conclusion is that a "self-energy matrix", $\hat{\Sigma}(u)$ or coherent medium, exists with a formal expression, Eq. (3.8), for its evaluation.

Now, if we take the s, s_0 matrix element of the algebraic identity

$$u(u\hat{1} - \hat{\Sigma})^{-1} = \hat{1} + \hat{\Sigma}(u\hat{1}-\hat{\Sigma})^{-1} , \tag{3.14}$$

we obtain the effective master equation

$$u<\tilde{P}(s,u|s_0)> = \delta(s,s_0) + \sum_{s'} \hat{\Sigma}_{ss'}<\tilde{P}(s',u|s_0)> \tag{3.15}$$

which is identical in form to Eq. (1.4) with the random $w_{ss'}$ replaced by the non-random $\hat{\Sigma}_{ss'}$.

Particle conservation in the form

$$\sum_{s} \tilde{P}(s,u|s_0) = 1/u \tag{3.16}$$

must remain true for the averaged quantities. Thus,

$$\sum_{s} \hat{\Sigma}_{ss'}(u) = 0 , \tag{3.17}$$

just as $\sum_{s} \hat{H}_{ss'} = 0$.

Our coherent jump rate $\hat{\Sigma}_{ss'}$ must display translational invariance

$$\hat{\Sigma}_{ss'}(u) = \hat{\Sigma}_{s-s',0}(u) . \tag{3.18}$$

Therefore, Eq. (3.15) can be solved by introducing the Fourier transformed quantities:

$$<\tilde{P}(\mathbf{k},u)> = \sum_{s} \exp\{-i\mathbf{k}\cdot(\mathbf{s}-\mathbf{s}_0)\}<\tilde{P}(s,u|s_0)> , \tag{3.19}$$

$$\hat{\Sigma}(\mathbf{k},u) \doteq \sum_{s \neq s_0} \hat{\Sigma}_{ss_0}(u)\exp\{-i\mathbf{k}\cdot(\mathbf{s}-\mathbf{s}_0)\} . \tag{3.20}$$

The use of the conservation law, Eq. (3.14), yields

$$\sum_{s}\hat{\Sigma}_{ss_0}(u)\exp\{-i\mathbf{k}\cdot(\mathbf{s}-\mathbf{s}_0)\} = \hat{\Sigma}(\mathbf{k},u)-\hat{\Sigma}(0,u) , \tag{3.21}$$

and

$$u<\tilde{P}(\mathbf{k},u)> = 1 + [\hat{\Sigma}(\mathbf{k},u)-\hat{\Sigma}(0,u)]<\tilde{P}(\mathbf{k},u)> . \tag{3.22}$$

The solution of the equation can be inverse Fourier transformed to yield

$$<\tilde{P}(s,u|s_0)> = \frac{1}{N}\sum_{\mathbf{k}}\frac{\exp\{i\mathbf{k}\cdot(\mathbf{s}-\mathbf{s}_0)\}}{u+\hat{\Sigma}(0,u)-\hat{\Sigma}(\mathbf{k},u)} , \tag{3.23}$$

$$= \frac{1}{u+\hat{\Sigma}(0,u)} \frac{1}{N} \sum_{\mathbf{k}} \frac{\exp\{i\mathbf{k}\cdot(\mathbf{s}-\mathbf{s}_0)\}}{1-\{\hat{\Sigma}(\mathbf{k},u)/[u+\hat{\Sigma}(0,u)]\}} , \qquad (3.24)$$

N being the total number of sites. This last result has precisely the same form as the CTRW result (Eq. 25) with Eq. (20) in Ref. 9(a) if one makes the identification

$$\Lambda(\mathbf{k},u) = \frac{\hat{\Sigma}(\mathbf{k},u)}{u+\hat{\Sigma}(0,u)} . \qquad (3.25)$$

Note that the prefactor in Eq. (25) of Ref. 9(a) then reduces to

$$(1/u)\{1-\Lambda(0,u)\} = (1/u)u/[u+\hat{\Sigma}(0,u)] \qquad (3.26)$$

in agreement with that in Eq. (3.24).

In particular, if $\hat{\Sigma}(\mathbf{k},u)$ is known, one can calculate $\tilde{\psi}(\mathbf{s},u)$ from

$$\tilde{\psi}(\mathbf{s},u) = \frac{1}{N} \sum_{\mathbf{k}} \Lambda(\mathbf{k},u) e^{i\mathbf{k}\cdot\mathbf{s}} \qquad (3.27)$$

$$= \frac{\hat{\Sigma}_{s0}}{u+\hat{\Sigma}(0,u)} . \qquad (3.28)$$

Here, $\tilde{\psi}(\mathbf{s},u)$ is the Laplace transform of the probability density $\psi(\mathbf{s},t)$ that the time between hops is t and the displacement $\mathbf{s}$. Unfortunately, the expression Eq. (3.13) is purely formal, and Klafter and Silbey do not propose a better method of evaluating it than the procedure used by Scher and Lax of averaging over a distribution of nearest neighbor distances (Section 6).

4. A simple example of the exact coherent medium

A simple example of the exact coherent medium can be obtained from Eq. (3.13) if we take a limit of $u=\infty$ or the high frequency limit. In this limit, the coherent medium is given simply by

$$\hat{\Sigma}(\infty) = <\hat{H}> . \qquad (4.1)$$

For example, if the sites $\{\mathbf{s}\}$ form a regular array of lattice points and if $\hat{H}$ is given by

$$\hat{H}_{\mathbf{ss'}} = \begin{cases} w_{\mathbf{ss'}} & \text{if } \mathbf{s} \text{ and } \mathbf{s'} \text{ are nearest neighbors} \\ -\Gamma_{\mathbf{s}} = \sum_{\mathbf{s'}(\text{n.n.})} w_{\mathbf{s's}} & \text{if } \mathbf{s}=\mathbf{s'} \\ 0 & \text{otherwise} \end{cases} , \qquad (4.2)$$

then the exact coherent medium has the same structure as that of $\hat{H}$ with the random jump rate $w_{\mathbf{ss}}$ replaced by the coherent jump rate w_c, which is defined by

$$w_c = <w_{\mathbf{ss'}}> . \qquad (4.3)$$

Here, we have assumed that each jump rate is distributed independently.

5. The coherent medium approximation

If we neglect the difference between the "exciting" matrix $\hat{G}^j$ and the "average" matrix $\hat{G}_c$ in Eq. (2.6) and make the crude approximation

$$\hat{G}^j \approx \hat{G}_c , \tag{5.1}$$

the rigorous condition Eq. (2.10) reduces to the simpler condition

$$\sum_j < \hat{T}_j > = 0 . \tag{5.2}$$

If each sub-unit $\hat{V}_j$ in Eq. (2.5) has an independent distribution, Eq. (5.2) is further reducible to a single condition

$$< \hat{T}_j > = 0 , \tag{5.3}$$

which supplies a condition to determine the unknown coherent medium. This idea[12,13] has been widely used in the problems of the lattice vibration[18] and the electronic energy band[19] in alloys and in other areas,[16] and is referred to as the "coherent potential approximation" (CPA). We shall use the more general nomenclature, "coherent medium approximation" (CMA), because as we shall see in the following, the idea is useful even when there are no potentials in the problem at all.[10] [We treat the jump rate instead of the potential!]

For the sake of simplicity, we assume that the localized centers $\{s\}$ form a regular lattice and that the jump rate $w_{ss'}$ is zero unless s and s' are nearest neighbors. We expect to find a coherent medium $\hat{\Sigma}(u)$ which is described only by the nearest neighbor coherent jump rate $w_c(u)$:

$$\hat{\Sigma}_{ss'} = w_c(u) \quad (s,s') \text{ nearest neighbors}$$

$$= 0 \qquad \text{otherwise} \tag{5.4}$$

and

$$\hat{\Sigma}_{ss} = -zw_c(u) , \tag{5.5}$$

where z is the coordination number of the lattice. If we replace $\hat{H}_1$ in Eqs. (2.2)-(2.4) by the coherent matrix, the difference matrix $\hat{V} \equiv \hat{H} - \hat{\Sigma}$ is written as

$$\hat{V} = \sum_{(s,s')} \hat{V}(s,s') , \tag{5.6}$$

where the summation runs all nearest neighbor pairs and the submatrix $\hat{V}(s,s')$ of large dimension contains non-vanishing elements in the 2×2 submatrix $\bar{V}$, for example for a pair of sites 1 and 2,

$$\bar{V} = \begin{bmatrix} w_c(u) - w_{21} & w_{12} - w_c(u) \\ w_{21} - w_c(u) & w_c(u) - w_{12} \end{bmatrix} . \tag{5.7}$$

If we introduce the abbreviated notation

$$\bar{P}_{ss'} \equiv \{(u\hat{1} - \hat{\Sigma})^{-1}\}_{ss'} \equiv \bar{P}(s,u|s') \tag{5.8}$$

and the 2×2 matrix

$$\bar{P} = \begin{bmatrix} P_{11} & P_{12} \\ P_{21} & P_{22} \end{bmatrix} , \tag{5.9}$$

then the condition Eq. (5.3) to determine $w_c(u)$ can be written more explicitly as

$$< \bar{V}(\bar{1} - \bar{P}\bar{V})^{-1} > = 0 , \tag{5.10}$$

where $\bar{1}$ is the unit 2×2 matrix, and the average is taken over the distribution for w_{12} and w_{21}.

For simplicity, we specialize to the case where $w_{ss'} = w_{s's}$. Then, it is easy to see that the four conditions contained in Eq. (5.10) reduces to a single condition

$$\left\langle \frac{w_c(u) - w_{12}}{1 - 2(\bar{P}_{11} - \bar{P}_{12})(w_c(u) - w_{12})} \right\rangle = 0 , \tag{5.11}$$

or equivalently

$$\bar{P}_{11} - \bar{P}_{12} = \left\langle \frac{1}{(\bar{P}_{11} - \bar{P}_{12})^{-1} - 2w_c(u) + 2w_{12}} \right\rangle . \tag{5.12}$$

Because the coherent medium lattice is assume to be periodic, $\bar{P}$ can be evaluated in the usual way as we used in Section 3:

$$\bar{P}(\mathbf{s}, u | \mathbf{s}_0) = \frac{1}{N} \sum_\mathbf{k} \frac{\exp\{-i\mathbf{k}\cdot(\mathbf{s}-\mathbf{s}_0)\}}{u + zw_c(u)(1 - f_\mathbf{k})} , \tag{5.13}$$

where

$$f_\mathbf{k} = \frac{1}{z}\sum_\mathbf{s}' \exp\{i\mathbf{k}\cdot(\mathbf{s}-\mathbf{s}_0)\} \tag{5.14}$$

and the primed sum is taken over the nearest neighbors of $\mathbf{s}_0$. Equation (5.13) can be used to obtain the relation

$$\bar{P}_{11} - \bar{P}_{12} = (1 - u\bar{P}_{11})/zw_c(u) , \tag{5.15}$$

so that only $\bar{P}_{11}$ needs to be evaluated.

By inserting Eq. (5.13) into Eq. (1.9), the diffusion constant of our coherent medium is simply given by

$$D(\omega) = a^2 w_c(i\omega) . \tag{5.16}$$

Here, a is the lattice constant (which may be modified for non-orthogonal lattices).

General results for limiting cases in the coherent medium approximation can be readily obtained. At the high frequency limit, our condition Eq. (5.11) yields

$$w_c(\infty) = \langle w_{12} \rangle \tag{5.17}$$

in agreement with the exact result Eq. (4.3). That is, our CMA is exact at the high-frequency limit.

At the static limit $u=0$, Eq. (5.11) is reducible to

$$\frac{1}{zw_c(0)} = \left\langle \frac{1}{(z-2)w_c(0) + 2w_{12}} \right\rangle . \tag{5.18}$$

We may formally express the solution of Eq. (5.18) in terms of a continued fraction

$$w_c(0) = \cfrac{2}{z\left\langle \cfrac{1}{w_{12} + \cfrac{z-2}{z\left\langle \cfrac{1}{w_{12} + \cfrac{z-2}{\cdots}} \right\rangle}} \right\rangle} . \tag{5.19}$$

In particular, if $z=2$, that is for linear chains, the coherent jump rate at the dc limit is given by

$$\frac{1}{w_c(0)} = \left\langle \frac{1}{w_{12}} \right\rangle , \tag{5.20}$$

which agrees with the result given by Bernasconi et al[20] on the basis of a scaling hypothesis.

The application of the CMA will be given in Sections 8-14.

6. Other approximations

A few approximate methods have been proposed to solve the random walk equation (1.1) or its Laplace transform, Eq. (1.4) with random jump rates. We summarize briefly these methods.

We start from Eq. (2.1) and separate the Hamiltonian matrix into the purely diagonal part $\hat{\Gamma}$ and the purely off-diagonal part $\hat{W}$:

$$\hat{H} = \hat{\Gamma} + \hat{W} , \qquad (6.1)$$

where

$$\hat{\Gamma}_{ss'} = -\hat{\Gamma}_s \delta(s,s') , \qquad (6.2)$$

$$\hat{W}_{ss'} = w_{ss'}(1-\delta(s,s')) . \qquad (6.3)$$

Then, the random walk propagator $\hat{G}$ can be expanded into a series

$$\hat{G} = (u\hat{1}-\hat{\Gamma})^{-1}\{\hat{1} + \sum_{m=1} [\hat{W}(u\hat{1}-\hat{\Gamma})^{-1}]^m\} . \qquad (6.4)$$

Scher and Lax[9,21] proposed to evaluate the average $<\hat{G}>$ in a Hartree approximation by

$$<\hat{G}> = <(u\hat{1}-\hat{\Gamma})^{-1}>\{\hat{1} + \sum_{m=1} [<\hat{W}><(u\hat{1}-\hat{\Gamma})^{-1}>]^m\} . \qquad (6.5)$$

This procedure yields the coherent medium whose diagonal and off-diagonal matrix elements are given by a weighted average of Γ_s and $w_{ss'}$, respectively,

$$\hat{\Sigma}_{ss} = -\left\langle \frac{\Gamma_s}{u+\Gamma_s} \right\rangle / \left\langle \frac{1}{u+\Gamma_s} \right\rangle , \qquad (6.6)$$

and

$$\hat{\Sigma}_{ss'} = \left\langle \frac{w_{ss'}}{u+\Gamma_{s'}} \right\rangle / \left\langle \frac{1}{u+\Gamma_{s'}} \right\rangle . \qquad (6.7)$$

The function $\tilde{\psi}(s,u)$ can be obtained using Eq. (3.28). This method has been successfully used to investigate many experiments.[6,9]

If we consider the random walk on a lattice and treat the case in which the jump rates are non-zero only between nearest neighbor pairs, we can express matrix elements of $\hat{G}$ as a renormalized perturbation series expansion.[22]

The off-diagonal element of $\hat{G}$ is written as

$$\hat{G}_{ij} = \hat{G}_{ii} w_{ij} \hat{G}_{jj}^{(i)} + \sum_{k \neq i,j} \hat{G}_{ii} w_{ik} \hat{G}_{kk}^{(i)} w_{kj} \hat{G}_{jj}^{(i,k)} + \cdots , \qquad (6.8)$$

where the summation extends over all self-avoiding paths starting at i and ending at j, and the diagonal element is given by

$$\hat{G}_{ii} = (u - \Gamma_i - \Delta_i) , \qquad (6.9)$$

and

$$\hat{G}_{kk}^{(i)} = (u - \Gamma_k - \Delta_k^{(i)}) , \qquad (6.10)$$

and so on. The superscripts (i), (i,k), $\cdots$ denote that these sites must be excluded from the paths. Here, an abbreviation $i=s_i$ is used. The "self-energy" Δ_i is given by

$$\Delta_i = \sum_k w_{ik} \hat{G}_{kk}^{(i)} w_{ki} + \sum_{k \neq l \neq i} w_{ik} \hat{G}_{kk}^{(i)} w_{kl} \hat{G}_{ll}^{(i,k)} w_{li} + \cdots , \qquad (6.11)$$

where, again, the summations are over all closed self-avoiding walks starting at i and ending at i, and $\Delta_k^{(i)}$ is given by a similar series with an extra excluded site.

A simplification is achieved if one applies the expansion Eq. (6.8) to the Bethe lattice[23] where no closed loops occur in the lattice. Since an exclusion of a site in the Bethe lattice breaks the lattice into z semi-infinite branches (z being the coordination number), the self-energies are given by a finite sum

$$\Delta_i = \sum_j^z{}' w_{ij} \hat{G}_{jj}^{(i)} w_{ji} \;, \qquad (6.12)$$

and

$$\Delta_k^{(i)} = \sum_{l \neq i}^{z-1}{}' w_{kl} \hat{G}_{ll}^{(k)} w_{lk} \;. \qquad (6.13)$$

Here, use has been made of $\hat{G}_{ll}^{(i,k)} = \hat{G}_{ll}^{(k)}$ in the Bethe lattice. The primed sum again sums over nearest neighbors and in Eq. (6.13) one of the nearest neighbors of site k is excluded.

Now, let us introduce

$$g_i^{(j)} \equiv w_{ij} - w_{ij} \hat{G}_{jj}^{(i)} w_{ji} \;. \qquad (6.14)$$

Then, the diagonal element of $\hat{G}$ is reduced to

$$\hat{G}_{ii} = \left[u + \sum_j^z{}' g_i^{(j)} \right]^{-1} \;, \qquad (6.15)$$

and $g_i^{(j)}$ satisfies

$$\frac{1}{g_i^{(j)}} = \frac{1}{w_{ij}} + \frac{1}{u + \sum_{k \neq i}^{z-1}{}' g_j^{(k)}} \;. \qquad (6.16)$$

Here, we have assumed $w_{ij} = w_{ji}$. Equations (6.15) and (6.16) imply that $\hat{G}_{ii}$ is expressed in terms of a continued fraction:

$$\hat{G}_{ii} = \cfrac{1}{u + \sum_j^z{}' \cfrac{1}{\cfrac{1}{w_{ij}} + \cfrac{1}{u + \sum_{k \neq i}^{z-1}{}' \cfrac{1}{\cfrac{1}{w_{jk}} + \cdots}}}} \;. \qquad (6.17)$$

Since a one-dimensional chain is a Bethe lattice with $z=2$, the expansions Eqs. (6.15)-(6.17) are valid for chains.

Bernasconi et al[24] examined the distribution function of the continued fraction $g_i^{(j)}$ for chains and proposed an approximate procedure to evaluate the effective continued fraction $g_c(u)$ by

$$g_c(u) = \left\langle \left[\frac{1}{w_{ij}} + \frac{1}{u + g_c(u)} \right]^{-1} \right\rangle \;. \qquad (6.18)$$

From the effective continued fraction, we can obtain the coherent medium with the coherent jump rate $w_c(u)$ given by

$$w_c(u) = \{u + g_c(u)\} g_c(u)/u \;. \qquad (6.19)$$

Movaghar[25] applied the same idea to Eq. (6.18) for the Bethe lattice with $z > 2$. However, the definition of the diffusion constant in the Bethe lattice is ambiguous. Movaghar assumed a completely random orientation of branches as in random branching high-polymers.

A comparison of these methods with an exact result will be given in Section 10.

7. Bond-percolation model in a chain I - Exact solution

The bond-percolation model is characterized by the following special type of distribution function $P(w_{ss'})$ for the nearest neighbor jump rate $w_{ss'}=w_{s's}$

$$P(w_{ss'}) = p\,\delta(w_{ss'}-w_0) + q\,\delta(w_{ss'}) \tag{7.1}$$

with $p+q=1$. Here, w_0 is a non-zero constant. This model is a realization of random interruptions in actual systems. The random distribution of the broken bonds decomposes the chain into a set of chunks. It is manifest that $\tilde{P}(\mathbf{s},u\,|\,\mathbf{s}_0)$ in the definition of $D(\omega)$ Eq. (1.9) is zero unless $\mathbf{s}$ belongs to the chunk which contains $\mathbf{s}_0$. An average treating all $\mathbf{s}_0$ as equally likely[26] yields

$$D(\omega) = \sum_{N=1}^{\infty} N(1-p)^2\, p^{N-1}\, D_N(\omega)\;, \tag{7.2}$$

where

$$D_N(\omega) = -\frac{\omega^2}{2N}\sum_{\mathbf{s},\mathbf{s}_0}^{N}(\mathbf{s}-\mathbf{s}_0)^2\,\tilde{P}(\mathbf{s},i\omega\,|\,\mathbf{s}_0) \tag{7.3}$$

is the diffusion constant of a chunk with N sites. The prefactor of the summand in Eq. (7.2)

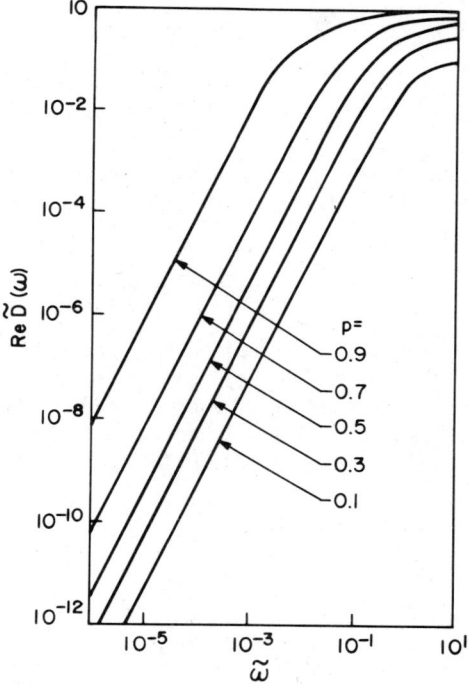

Fig. 7.1. The dependence of the real part of $\tilde{D}(\omega)$ on frequency for various values of p. The exact solution for the bond-percolation model in a chain. (Taken from ref. 27.)

Fig. 7.2. The dependence of the imaginary part of $\tilde{D}(\omega)$ on frequency for various values of p. The exact solution for the bond-percolation model in a chain. (Taken from ref. 27.)

is the probability that a given site belongs to a chunk with N sites.

Now, the conductivity for the finite chunk Eq. (7.3) can be obtained by diagonalizing an $N \times N$ submatrix of $\hat{H}$ to yield[27]

$$D_N(\omega) = a^2 w_0 \left[1 + \frac{(1+4/i\tilde{\omega})^{1/2}}{N} \left\{ \frac{1}{z_+^{2N}+1} - \frac{1}{z_-^{2N}+1} \right\} \right], \quad (7.4)$$

where $\tilde{\omega} = \omega/w_0$ and $z_\pm = (\sqrt{i\tilde{\omega}} \pm \sqrt{4+i\tilde{\omega}})/2$. The frequency dependent diffusion constant can be easily evaluated from Eq. (7.2). Figures 7.1 and 7.2 show the frequency dependence of the real and imaginary part of the dimensionless diffusion constant $\tilde{D}(\omega) = D(\omega)/a^2 w_0$. We can show that the real and imaginary parts of $D(\omega)$ and hence of the conductivity vanish quadratically and linearly, respectively, with frequency. In fact, we have

$$A \equiv \lim_{\tilde{\omega} \to 0} \operatorname{Re}\tilde{D}(\omega)/\tilde{\omega}^2 = \frac{p(1+p)^2}{4(1-p)^4} \quad (7.5)$$

and

$$B \equiv \lim_{\tilde{\omega} \to 0} \operatorname{Im}\tilde{D}(\omega)/\tilde{\omega} = \frac{p}{2(1-p)^2}, \quad (7.6)$$

which are plotted in Fig. 7.3. The quantities A and B diverge at the percolation threshold $p=1$.

The exact diffusion constant may be used to determine an exact coherent jump rate between nearest neighbors through $w_c(\omega) = D(\omega)/a^2$.

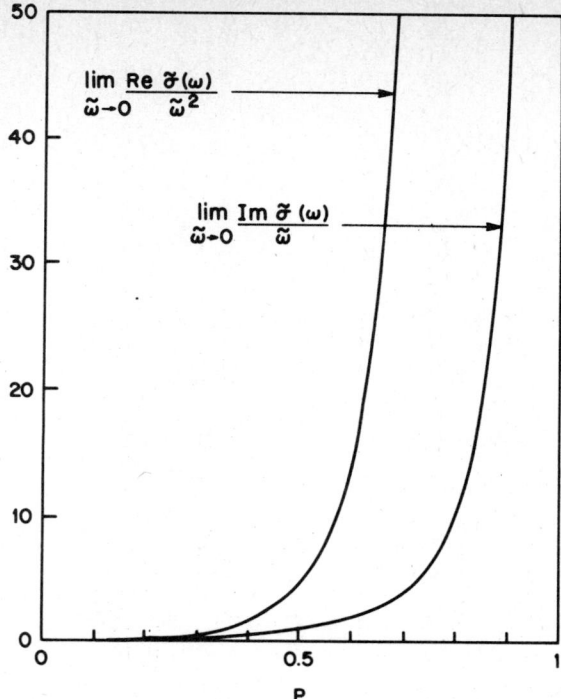

Fig. 7.3. Critical behavior of $\lim_{\tilde{\omega}\to 0} \mathrm{Re}\,\tilde{D}(\omega)/\tilde{\omega}^2$ and $\lim_{\tilde{\omega}\to 0} \mathrm{Im}\,\tilde{D}(\omega)/\tilde{\omega}$ at percolation threshold $p=1$. The exact solution for the bond-percolation model in a chain. Explicit forms are given in Eqs. (7.5) and (7.6).

8. The coherent medium approximation in one-dimensional chains

For one-dimensional chains, the diagonal element of the coherent random walk propagator $\bar{P}_{11}$ can be easily evaluated using Eq. (5.13) with $\mathbf{s}=\mathbf{s}_0$:

$$\bar{P}_{11} \equiv \bar{P}(\mathbf{s}_0, u \mid \mathbf{s}_0) = \{u(u+4w_c)\}^{-1/2} . \tag{8.1}$$

Noting Eq. (5.15), we can discuss general features of the ac conductivity on the basis of the CMA condition Eq. (5.12).

The high frequency behavior of the ac conductivity is determined by the first and second moment of the jump rate:

$$\tilde{\sigma}(\omega) \approx \langle w_{12}\rangle/w_0 + 2i\{\langle w_{12}^2\rangle - \langle w_{12}\rangle^2\}/w_0^2\tilde{\omega} , \tag{8.2}$$

where w_0 is a relevant scaling parameter of the frequency and $\tilde{\omega} \equiv \omega/w_0$.

The low-frequency behavior of the conductivity is also obtained using an expansion for small u in Eq. (5.11) with Eqs. (5.15) and (8.1). We have found the following six regimes which are distinguished by the values of the first, $m_1 \equiv w_0\langle 1/w_{12}\rangle$, and second, $m_2 \equiv w_0^2\langle 1/w_{12}^2\rangle$, moments of the inverse of the jump rate[28]:

(1) $m_1, m_2 < \infty$:

$$\tilde{\sigma}(\omega) \approx m_1^{-1} + \{(m_2/m_1^2 - 1)/2m_1^{1/2}\}(i\tilde{\omega})^{1/2} . \tag{8.3}$$

(2) $m_1<\infty, m_2=\infty$ (logarithmic) with a distribution function $P(w_{12}) \propto w_{12}$:

$$\tilde{\sigma}(\omega) \approx m_1^{-1} - C_1 m_1^{-5/2}(i\tilde{\omega})^{1/2}\ln(i\tilde{\omega})^{1/2} . \tag{8.4}$$

(3) $m_1<\infty, m_2=\infty$ (non-logarithmic) with a distribution function $P(w_{12}) \propto (w_{12})^\alpha (0<\alpha<1)$:

$$\tilde{\sigma}(\omega) \approx m_1^{-1} + C_2 m_1^{-(4+\alpha)/2} \cdot (i\tilde{\omega})^{\alpha/2} \tag{8.5}$$

(4) $m_1=\infty$ (logarithmic), $m_2=\infty$ with a distribution function $P(w_{12}) \propto$ constant:

$$\tilde{\sigma}(\omega) \approx -C_3/\ln(i\tilde{\omega}) . \tag{8.6}$$

(5) $m_1=\infty$, $m_2=\infty$ with a distribution function $P(w_{12}) \propto (w_{12})^\alpha$ $(-1<\alpha<0)$:

$$\tilde{\sigma}(\omega) \approx C_4(i\tilde{\omega})^{-\alpha/(2+\alpha)} . \tag{8.7}$$

(6) $m_1=\infty$, $m_2=\infty$ with a distribution function $P(w_{12}) \propto q\delta(w_{12}) + pF(w_{12})$ $(p+q=1)$, $\delta(w_{12})$ is a Dirac δ function and $F(w_{12})$ is any nonsingular function):

$$\tilde{\sigma}(\omega) \approx \{(1-q^2)/4q^2\}(i\tilde{\omega}) . \tag{8.8}$$

Here, $C_1,...,C_4$ are constants.

Using the approximation Eq. (6.18)[24,29] and a scaling argument on the distribution function of the continued fraction Eq. (6.16), Bernasconi et al[20] discussed general classes of the distribution function of the jump rate on the basis of the asymptotic behavior of $<P(s_0,u|s_0)>$ near $u=0$. Their results are summarized in a recent review article.[30] They classified Eq. (8.3)-(8.5) as class (a), and Eq. (8.6) as class (b) and Eq. (8.7) as class (c).

The conditional probability that a carrier remains at its initial site after an infinite time is given by

$$P_{00}(\infty) \equiv \lim_{t\to\infty} <P(s_0,t|s_0,0)> \tag{8.9}$$

$$= \lim_{u\to 0} u\, \bar{P}(s_0,u|s_0) . \tag{8.10}$$

Using the coherent jump rate in Eq. (8.1), we can easily verify that $P_{00}(\infty)=0$ for cases (1)-(5) and $P_{00}(\infty)=q$ for case (6). Therefore, cases (4) and (5) supply examples in which a carrier can diffuse away infinitely from its initial position even though the dc conductivity vanishes.[25,28,30]

9. Binary jump rate in a chain

Suppose a chain whose nearest neighbor jump rate is distributed according to a function

$$P(w_{12}) = p\, \delta(w_{12}-w_0) + q\, \delta(w_{12}-w_0\delta) , \tag{9.1}$$

where $p+q=1$, $w_0 \neq 0$ and $0<\delta \leqslant 1$. ($\delta=0$ corresponds to the bond-percolation model, whose exact result has been given in Section 7. The CMA treatment of this case will be given in the next section.) It is clear that the first and the second moments of the inverse of the jump rate are finite:

Fig. 9.1. The dependence on p and δ of $\tilde{\sigma}(\infty)$ and $\tilde{\sigma}(0)$ of a chain with a binary jump rate.

Fig. 9.2. The frequency dependence of the real part of $\tilde{\sigma}(\omega)$ for various values of p: binary jump in a chain. $\delta=0$ corresponds to the bond-percolation model.

$$m_1 = w_0 <1/w_{12}> = (p\delta+q)/\delta , \quad (9.2)$$

and

$$m_2 = w_0^2 <1/w_{12}^2> = (p\delta^2+q)/\delta^2 . \quad (9.3)$$

Therefore, this distribution belongs to case (1) in Section 8, and $\tilde{\sigma}(0)$ is given by $1/m_1=\delta/(q+p\delta)$. Since the conductivity at $\omega=\infty$ is given by Eq. (8.2)

$$\tilde{\sigma}(\infty) = <w_{12}>/w_0 = p+q\delta , \quad (9.4)$$

the ac conductivity for fixed p and δ shows a dispersive frequency dependence between these limiting values $\tilde{\sigma}(0)$ and $\tilde{\sigma}(\infty)$ which are plotted in Fig. 9.1. Actual frequency dependence can be readily evaluated using Eqs. (5.11), (5.15), (8.1) and (5.16). Figure 9.2 shows the frequency dependence of the real part of $\tilde{\sigma}(\omega)$ for $p=0.9, 0.5, 0.1$. The ac part $\tilde{\sigma}(\omega)-\tilde{\sigma}(0)$ is given by

$$\tilde{\sigma}(\omega) - \tilde{\sigma}(0) = \frac{pq\sqrt{\delta}(1-\delta)^2}{2\sqrt{2}(q+p\delta)^{5/2}} (1+i)\omega^{1/2} , \quad (9.5)$$

near $\omega=0$. The prefactor is shown in Fig. 9.3 as a function of δ for $p=0.9, 0.5$ and 0.1.

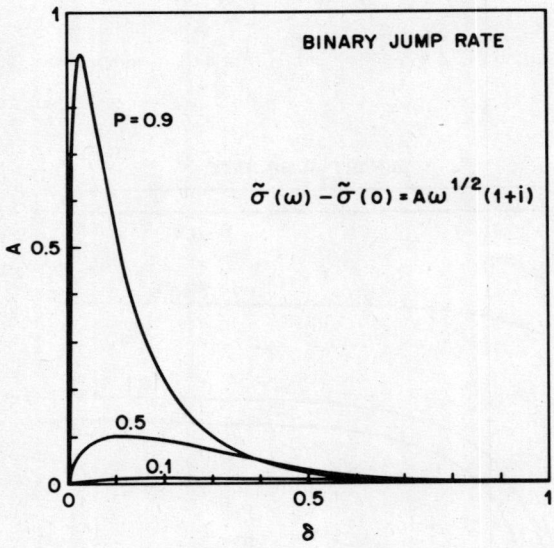

Fig. 9.3. Plots of the prefactor on the right-hand side of Eq. (9.5). The prefactor vanishes at $\delta=0$ and $\delta=1$ as it should be.

10. Bond-percolation model in a chain II. - Approximate treatment

We apply the approximate methods discussed in Section 5 and Section 6 to the bond-percolation model in a chain to provide a comparison of their relative merits.[26]

If we use the distribution function Eq. (7.1) in Eq. (5.11) for the CMA,[10] in Eq. (6.6) for

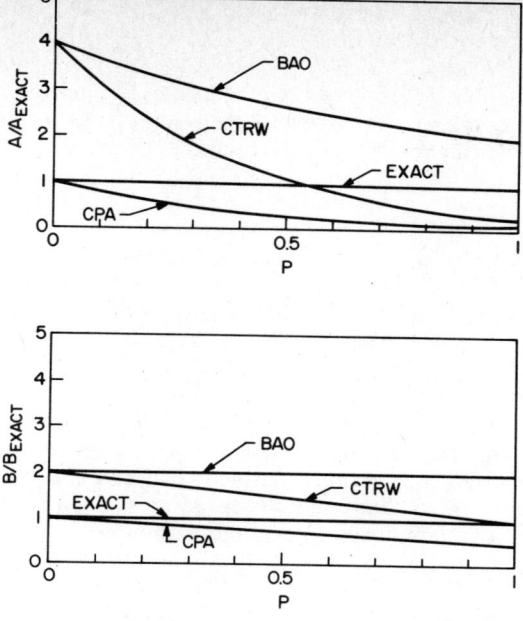

Fig. 10.1. A comparison of the relative merits of the various approximations in the low-frequency regime for the bond-percolation model in a chain: (1) Exact, (2) CMA [Eq. (5.11)], (3) SL [Eq. (6.6)], (4) BAO [Eq. (6.18)]. (Taken from ref. 27.)

the Scher-Lax approximation (SL)[9] and in Eq. (6.18) for the approximation due to Bernasconi et al (BAO),[24] we obtain the coherent jump rate for each method: for the CMA[10]

$$w_c(u)/w_0 = [\tilde{u} + 2(1-p)^2 - (1-p)\{(\tilde{u}+2)^2 + 4p(p-2)\}^{1/2}]/\tilde{u} , \quad (10.1)$$

from Eq. (6.6)

$$w_c(u)/w_0 = p(2-p+\tilde{u})\tilde{u}/[2(1-p)^2 + (3-2p)\tilde{u} + \tilde{u}^2] \quad (10.2)$$

and from Eq. (6.18)

Table I. The low frequency behavior of $\tilde{\sigma}(\omega)$

Method	A	B
Exact	$p(1+p)^2/4(1-p)^4$	$p/2(1-p)^2$
CMA	$p(2-p)/8(1-p)^4$	$p(2-p)/4(1-p)^2$
SL	$p(4-3p)/4(1-p)^4$	$p(2-p)/2(1-p)^2$
BAO	$p(1+p)/(1-p)^4$	$p/(1-p)^2$

$$w_0(u)/w_0 = [(1-p)^2 + (1+p)u - (1-p)\{(1-p)^2 + 2(1+p)\tilde{u} + \tilde{u}^2\}^{1/2}]/2\tilde{u} \ . \tag{10.3}$$

Although we can compare the conductivity in the entire range of the frequency, we supply here only a comparison of coefficients A and B defined by Eqs. (7.5) and (7.6) in Table I and Fig. 10.1. In the latter, the ratio of A/A_{exact} and B/B_{exact} are plotted versus p. We would like to mention that Bernasconi et al believe their approximation Eq. (6.18) may not be adequate for the case in which completely broken jump rates are possible.[26,29,30]

11. Chains with positionally random localized centers

Let us consider a chain in which hopping centers are distributed randomly in space and the jump rate $w_{l,m}$ of a carrier between m and l hopping sites obeys

$$w_{l,m} = w_0 \exp(-|x_m - x_l|/R_d) \ . \tag{11.1}$$

Here, R_d is half an effective Bohr radius of the localized carrier and x_m denotes the position of hopping site m. We assume that $w_{l,m}$ is negligible unless m and l are adjacent hopping sites, because the carrier hopping will be most frequent between adjacent sites. As is well known, the probability density $N(x)$ that one finds an adjacent hopping site at a distance x from a given site in a uniform system obeys a Poisson distribution

$$N(x) = n_s \exp(-n_s x) \ , \tag{11.2}$$

where n_s is the number density of the hopping sites. Therefore, the jump rate w between adjacent centers is distributed according to

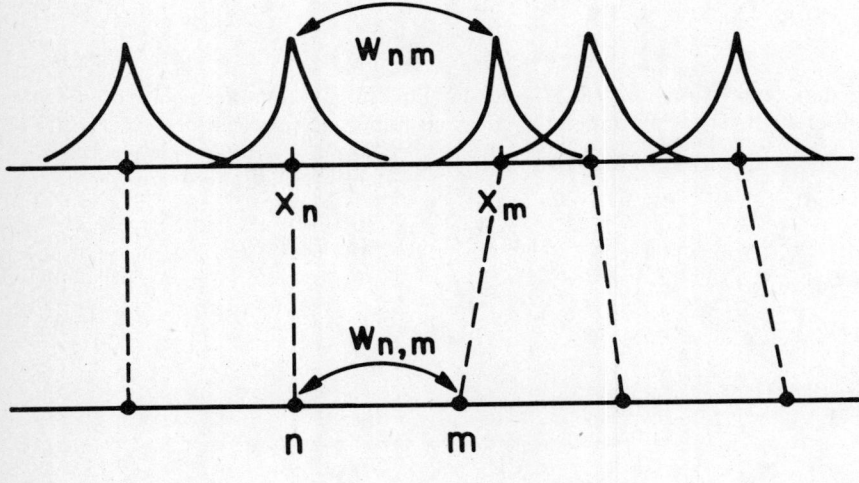

Fig. 11.1. Schematic illustration of mapping between the actual system and the lattice model. The jump rate between two adjacent centers is given by Eq. (11.1), which is mapped into w_{nm} on a regular chain. The latter obeys the distribution Eq. (11.3). (Taken from ref. 31.)

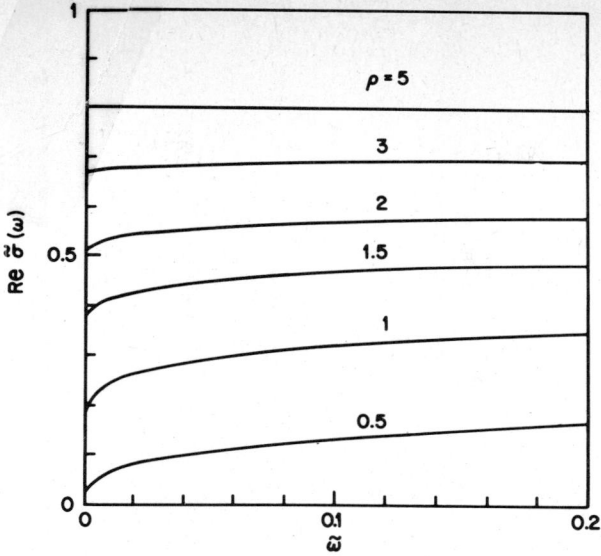

Fig. 11.2. The frequency dependence of the real part of $\tilde{\sigma}(\omega)$: a power law distribution in a chain. (Taken from ref. 31.)

$$P(w) = \begin{cases} \rho w^{\rho-1}/w_0^\rho & 0 \leqslant w \leqslant w_0 \\ 0 & \text{otherwise} \end{cases}, \qquad (11.3)$$

where $\rho = n_s R_d$ is the dimensionless density of hopping sites.

Now, we map the actual positionally random system onto a regular lattice whose lattice constant is $a \equiv 1/n_s$ and the jump rate between nearest neighbors obeys the distribution Eq. (11.3). A schematic correspondence of the mapping is shown in Fig. 11.1.[31]

The distribution function Eq. (11.3) has already been used in Section 8 as a prototype function of the general distribution. Using the CMA approximation Eq. (5.11) with Eqs. (5.15) and (8.1), we obtain the frequency dependence of the ac conductivity as shown in Fig. 11.2.[31] As we decrease the density ρ from infinity to zero, the five regimes (1)-(5) given in Section 8 appear one by one. Explicitly, the low-frequency behavior of the conductivity is shown to have the following form:

(1) $\rho > 2$

$$\tilde{\sigma}(\omega) \approx \frac{\rho-1}{\rho} + \frac{1}{2\rho(\rho-2)} \left[\frac{\rho-1}{\rho}\right]^{1/2} (i\omega)^{1/2} , \qquad (11.4)$$

(2) $\rho = 2$

$$\tilde{\sigma}(\omega) \approx \frac{1}{2} - \frac{1}{4\sqrt{2}} (i\tilde{\omega})^{1/2} \ln(i\tilde{\omega})^{1/2} , \qquad (11.5)$$

(3) $1 < \rho < 2$

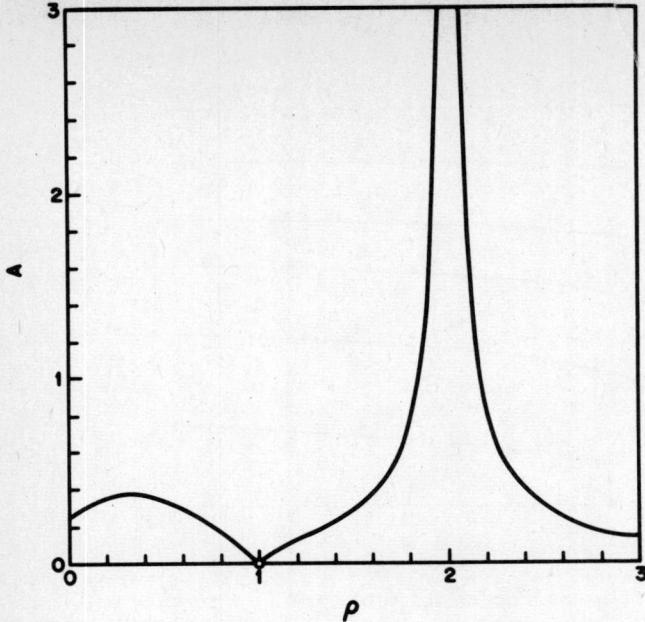

Fig. 11.3. The dependence of factor A on ρ. Open circle denotes a singular point. (Taken from ref. 31.)

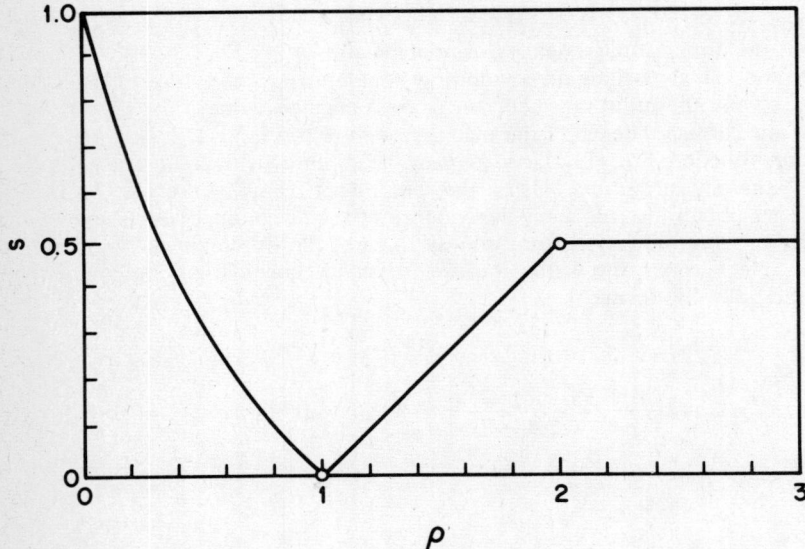

Fig. 11.4. The dependence of index s on ρ. Open circles denote singular points. (Taken from ref. 31.)

$$\tilde{\sigma}(\omega) \approx \frac{\rho-1}{\rho} + 2^{1-\rho}\left[\frac{\rho-1}{\rho}\right]^{(\rho+1)/2} \frac{(\rho-1)\pi}{\sin(\rho-1)\pi} (i\omega)^{(\rho-1)/2} , \qquad (11.6)$$

(4) $\rho = 1$

$$\tilde{\sigma}(\omega) \approx -\frac{2}{\ln\{-i\tilde{\omega}/\ln(i\tilde{\omega})\}} , \qquad (11.7)$$

(5) $0 < \rho < 1$

$$\tilde{\sigma}(\omega) \approx \left\{\frac{2^{1-\rho}\rho\pi}{\sin\rho\pi}\right\}^{-\frac{2}{1+\rho}} (i\omega)^{\frac{1-\rho}{1+\rho}} . \qquad (11.8)$$

The ac part $\tilde{\sigma}(\omega)-\tilde{\sigma}(0)$ except for cases (2) and (4) has a form $A(i\omega)^s$. Figures 11.3 and 11.4 show the dependence of A and s on ρ, respectively. We can see that the ac part shows approximately symmetric behavior near $\rho=1$, which has been numerically observed by Richards and Renken.[32]

12. Bond-percolation model in a simple cubic lattice

We apply the CMA method to the bond percolation model in a simple cubic lattice. In order to obtain simple analytic results, we assume that $f_\mathbf{k}$ defined in Eq. (5.14) for the simple cublic lattice has a semi-elliptic density of states

$$\frac{1}{N}\sum_\mathbf{k} \delta(\epsilon-f_\mathbf{k}) = 2(1-\epsilon^2)^{1/2}/\pi , \qquad (12.1)$$

with the correct behavior at the band edges. Using Eq. (12.1), the diagonal element of the random walk propagator Eq. (5.13) with $\mathbf{s}=\mathbf{s}_0$ is written as

$$\bar{P}(\mathbf{s},u\,|\,\mathbf{s}) = 2\{u+zw_c+\sqrt{u(u+2zw_c)}\}^{-1} . \qquad (12.2)$$

The self-consistency equation (5.11) with Eqs. (7.1) and (12.2) can be solved analytically.[10] Figure 12.1 shows the frequency dependence of $\tilde{\sigma}(\omega)$. The dc conductivity vanishes at the percolation threshold p_c as

$$\tilde{\sigma}(0) = \begin{cases} \frac{3}{2}(p-p_c) & p \geqslant p_c \\ 0 & p \leqslant p_c \end{cases}, \qquad (12.3)$$

where $p_c=1/3$. The ac part of the conductivity shows the following critical behavior at the percolation threshold:

$$\lim_{\tilde{\omega}\to 0} \frac{\text{Re}\{\tilde{\sigma}(\omega)-\tilde{\sigma}(0)\}}{\tilde{\omega}^{3/2}} \sim (p-p_c)^{-3/2} \quad (p>p_c) , \qquad (12.4)$$

$$\lim_{\tilde{\omega}\to 0} \frac{\text{Re}\tilde{\sigma}(\omega)}{\tilde{\omega}^2} \sim (p-p_c)^{-3} \quad (p<p_c) \qquad (12.5)$$

and

$$\lim_{\tilde{\omega}\to 0} \frac{\text{Im}\tilde{\sigma}(\omega)}{\tilde{\omega}} \sim |p-p_c|^{-1} . \qquad (12.6)$$

At $p=p_c$, $\tilde{\sigma}(\omega)$ vanishes as $(i\omega)^{1/2}$. Therefore, both $<P(\mathbf{s}_0,\infty\,|\,\mathbf{s}_0,0)>$ and $\tilde{\sigma}(0)$ vanish at

Fig. 12.1. The frequency and the probability dependence of the real and imaginary parts of the dimensionless ac conductivity $\tilde{\sigma}(\omega)$ for the bond-percolation model in the simple cubic lattice. (Taken from ref. 10.)

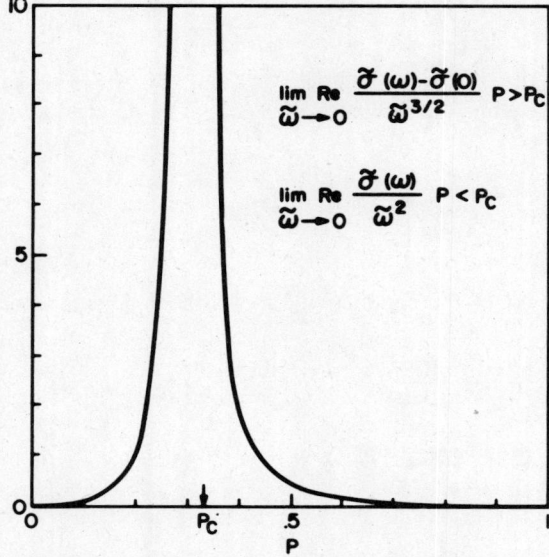

Fig. 12.2. The critical behavior of the real part of $\tilde{\sigma}(\omega)$ for the bond-percolation model in the simple cubic lattice at the static limit. [Equations (12.4) and (12.5).]

Fig. 12.3. The critical behavior of the imaginary part of $\tilde{\sigma}(\omega)$ for the bond-percolation model in the simple cubic lattice at the static limit. [(Eq. 12.6).]

the percolation threshold. The critical behavior of Eqs. (12.4)-(12.6) are depicted in Figs. 12.2 and 12.3. It should be remarked that the critical indices in Eqs. (12.3)-(12.6) are not changed even if we use the exact expression for $\bar{P}(s,u\,|\,s)$ instead of the approximation Eq. (12.2), although the proportional constants in Eqs. (12.4)-(12.6) are modified.[10]

13. Comparison with a computer simulation - Miller-Abrahams jump rate

In hopping between majority sites in doped semiconductors, Miller and Abrahams[33] derived the jump rate in the form

$$w_{ss'} \equiv w(r) = w_0(r/R_d)^{3/2} \exp(-r/R_d) , \qquad (13.1)$$

where $r = |\mathbf{s}-\mathbf{s}'|$ and w_0 is a constant dependent on the temperature. Here, we have assumed that the activation energy is independent of r. Instead of treating the actual positionally disordered system we map the system onto a simple cubic lattice where the nearest neighbor jump rate obeys a distribution which is implied by the jump rate (13.1) and the distribution function $N(r)$ of the nearest neighbor distance r in a random system of uniform average density N_D. It is known that $N(r)$ is given by the Hertz distribution[34]:

$$N(r) = \frac{4\pi r^2 N_D}{z} \exp\left\{-\frac{4\pi N_D}{3z} r^3\right\} , \qquad (13.2)$$

where $z(=6)$ is the coordination number of the simple cubic lattice and N_D is the density of hopping sites. The lattice constant a is chosen to be $(4\pi/3N_D)^{1/3}$. The CMA condition Eq. (5.11) reduces to

Fig. 13.1. Typical frequency dependence of the real and imaginary part of $\tilde{\sigma}(\omega)$ for the Miller-Abrahams jump rate (13.1). $\eta \equiv 4\pi N_D R_d^3 = 10^{-3}$.

Fig. 13.2. A comparison of the frequency dependence of the real part of the dimensionless conductivity $\tilde{\sigma}(\omega)$ for the Miller-Abrahams jump rate. The solid line is the result given by the present coherent medium approximation and the solid circles are the result of the computer simulation given by McInnes et al (ref. 35). $\eta \equiv 4\pi N_D R_d^3 = 2.15 \times 10^{-3}$.

$$\int_0^\infty \frac{N(r)\{w_c - w(r)\}\, dr}{1 - 2(\bar{P}_{11} - \bar{P}_{12})\{w_c - w(r)\}} = 0 \,. \tag{13.3}$$

Typical results are shown in Fig. 13.1 for the frequency dependence of the real and imaginary part of the ac conductivity, where Eq. (12.2) has been used for $\bar{P}_{11}$ and the set of self-consistency equations (5.15), (12.2), (13.1), (13.2) and (13.3) has been solved numerically.

J. A. McInnes et al[35] performed a computer simulation on the positionally disordered system. A comparison is given in Fig. 13.2. The agreement between the CMA method and their computer simulation is good in the low-frequency region. The discrepancy at the high-frequency may be attributed to our use of the lattice model.

14. Impurity conduction

For impurity conduction which we have explained in Section 1, Scher and Lax[9] have shown that a simplified jump rate

$$w(r) = w_0 \exp(-r/R_d) \,, \tag{14.1}$$

describes experiments very well. We have used the same notation w_0 for the coefficient in Eq. (14.1), which takes a different value from w_0 in Eq. (13.1). Using the same method as in Section 13, we obtained the ac conductivity for doping levels $N_D = 2.7 \times 10^{17} \text{cm}^{-3}$ and $1.2 \times 10^{16} \text{cm}^{-3}$.[10] Figures 14.1 and 14.2 show a comparison with experiments given by Pollak

Fig. 14.1. Comparison of the theoretical and experimental values of Re $\tilde{D}(\omega)$ for $N_D = 2.7 \times 10^{17} \text{cm}^{-3}$. The solid line is the present result and the broken line is the SL result. The latter and experimental data for Re $\sigma(\omega)$ are taken from Ref. 9(b). The dimensionless ac conductivity or diffusion constant $\tilde{D}(\omega) = \sigma(\omega)/[(ne^2/kT)(a^2 w_0)]$ with $n = 0.8 \times 10^{15} \text{cm}^{-3}$ and $a = (4\pi N_D/3)^{-1/3}$ is shown. (Taken from ref. 10.)

Fig. 14.2. Comparison of the theoretical and experimental values of $\text{Re}\,\tilde{D}(\omega)$ for $N_D = 1.2 \times 10^{16} \text{cm}^{-3}$. Symbols are the same as in Fig. 14.1. According to SL, a is assumed to be $0.9\,(4\pi N_D/3)^{-1/3}$ in this plot. (Taken from ref. 10.)

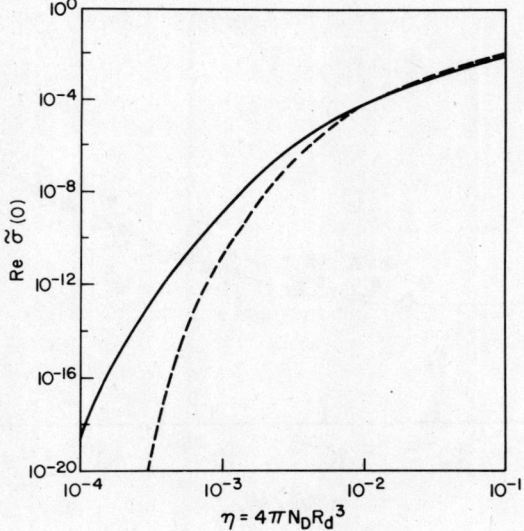

Fig. 14.3. Comparison of the present method (solid line) and the SL method (broken line) for the static dimensionless diffusion constant. $\text{Re}\,\sigma(0)/a^2 w_0 = \frac{1}{2}\exp\{-(18\ln 3/2)^{1/3}\eta^{-1/3}\}$ for the present method and $e^\gamma \eta^{1/4}\exp\{-2\eta^{-1/2}/3\}/6\sqrt{\pi}$ for the SL method with $\eta = 4\pi N_D R_d^3$ and $\gamma = 0.5772$. (Taken from ref. 10.)

and Geballe.[1] The results given by Scher and Lax[9] (SL) using the CTRW method are also shown in these figures. The ac part of our results shows the well-known ω^s dependence and the exponent s reads as 0.52 for three decades of $\tilde{\omega}$ in Fig. 14.1 and 0.77 for seven decades of $\tilde{\omega}$ in Fig. 14.2.

At low densities, the static conductivity can be shown to have the density dependence

$$\tilde{\sigma}(0) \approx \frac{1}{2} \exp\{-(18\ln 3/2)^{1/3} \eta^{-1/3}\}, \quad (14.2)$$

where $\eta = 4\pi N_D R_d^3$. A comparison between the CTRW and the CMA methods for $\tilde{\sigma}(0)$ is given in Fig. 14.3.

15. Summary

We have presented the coherent medium approach to hopping conduction problems where the motion of carriers obey the usual random walk equation and discussed the existence of the coherent medium which is defined through an average of the random walk propagator. We have introduced the coherent medium approximation (CMA) to obtain an approximate but easily tractable coherent medium. The CMA is a generalized use of the coherent potential approximation (CPA) in the master equation. The CPA is one of the most fruitful methods in treating random systems and is widely used in electron[19] and phonon problems.[18] We have applied the CMA to various cases of hopping conduction in one- and three-dimensions and compared the results with experiment. A simple comparison of the CMA with other methods has also been given.

We have derived the CMA condition Eq. (5.11) from the multiple scattering formalism due to Lax.[12,13] An alternative derivation of Eq. (5.11) was given by Odagaki and Lax,[10] where a traditional idea of the effective medium approximation was used.[36] Namely, a random unit is subjected to an as yet unknown effective medium and the effective medium is determined to be such that the resulting extra perturbation is required to vanish on the average over all possibilities of the random unit. A similar condition to Eq. (5.11) has also been used in the problem of random resistor networks.[37]

We hope that further improvements of the coherent medium approximation will be developed including traps, asymmetric jump rates, cluster effects and a similar method will be applied to other problems, for example, optical properties and magnetic properties of random systems.

Acknowledgements

The work at City College was supported in part by a grant from the Army Research Office, the PSC-CUNY Research Award Program, Department of Energy, and in conjunction with the binational agreement between the National Science Foundation and the Japanese Society for the Promotion of Science, under contract No. INT-7918591.

References

[1] M. Pollak and T. H. Geballe, Phys. Rev. **122**, 1742 (1961).
[2] J. M. Reyes, M. Sayer, A. Mansingh and R. Chen, Can. J. Phys. **54**, 413 (1976).
[3] A. S. Barker Jr., J. A. Ditzenberger and J. P. Remeika, Phys. Rev. **B14**, 4254 (1976).
[4] M. Sayer, A. Mansingh, J. B. Webb and J. Noad, J. Phys. C: Solid State Phys. **11**, 315 (1978).
[5] R. M. Mehra, P. C. Mathur, A. K. Kathuria and R. Shyam, Phys. Rev. **B18**, 5620 (1978).

[6] M. Suzuki, J. Phys. Chem. Solids **41**, 1253 (1980).

[7] S. R. Elliott, Phil. Mag. **36**, 1291 (1977).

[8] Actually, $P(s,t|s_0,0)$ can be written as an absolute square of a matrix element of $e^{-iHt/\hbar}$, where H is the total Hamiltonian of the underlying problem. See refs. 9 and 10.

[9] H. Scher and M. Lax, (a) Phys. Rev. **B7**, 4491 (1973); (b) ibid, 4502 (1973).

[10] T. Odagaki and M. Lax, Phys. Rev. B to be published.

[11] For example, T. Holstein, S. K. Lyo and R. Orbach, Phys. Rev. **B15**, 4693 (1977).

[12] M. Lax, Rev. Mod. Phys. **23**, 287 (1951); Phys. Rev. **85**, 621 (1952).

[13] M. Lax, "Wave Propagation and Conductivity in Random Media," in *Stochastic Differential Equations* SIAM-AMS (Soc. for Industrial and Applied Math. - American Mathematical Soc.) Proc. vol. **6**, 35-95 Amer. Math. Soc. Providence, R.I. (1973).

[14] J. Klafter and R. Silbey, Phys. Rev. Letts. **44**, 55 (1980).

[15] R. Zwanzig, J. Chem. Phys. **33**, 1338 (1960); See also *Lectures in Theoretical Physics* Vol. **III**, 106 (edited by W. E. Brittin), Interscience, New York (1961); and Phys. Rev. **124**, 983 (1961).

[16] See, for example, F. Yonezawa and K. Morigaki, Prog. Theor. Phys. Suppl. **53**, 1 (1973); R. J. Elliott, J. A. Krumhansl and P. L. Leath, Rev. Mod. Phys. **46**, 465 (1974).

[17] M. Lax, Phys. Rev. **79**, 200 (1950).

[18] D. W. Taylor, Phys. Rev. **156**, 1017 (1967).

[19] P. Soven, Phys. Rev. **156**, 809 (1967).

[20] J. Bernasconi, W. R. Schneider and W. Wyss, Z. Physik B **37**, 175 (1980).

[21] M. Lax and H. Scher, Phys. Rev. Letts. **39**, 781 (1977).

[22] E. Feenberg, Phys Rev. **74**, 206 (1948); E. N. Economou, "*Green's Functions in Quantum Physics*," (Springer-Verlag, Berlin, Heidelberg 1979).

[23] R. Abou-Chacra, P. W. Anderson, and D. J. Thouless, J. Phys. C: Solid State Phys. **6**, 1734 (1973).

[24] J. Bernasconi, S. Alexander and R. Orbach, Phys. Rev. Letts. **41**, 185 (1978).

[25] B. Movaghar, J. Phys. C: Solid State Phys. **13**, 4915 (1980).

[26] T. Odagaki and M. Lax, Phys. Rev. B to be published.

[27] T. Odagaki and M. Lax, Phys. Rev. Letts. **45**, 847 (1980).

[28] T. Odagaki and M. Lax, in preparation.

[29] J. Bernasconi, H. U. Beyeler, S. Strässler and S. Alexander, Phys. Rev. Letts. **42**, 819 (1979).

[30] S. Alexander, J. Bernasconi, W. R. Schneider and R. Orbach, Rev. Mod. Phys. **53**, 175 (1981).

[31] T. Odagaki and M. Lax, in preparation.

[32] P. M. Richards and R. L. Renken, Phys. Rev. **B21**, 3740 (1980).

[33] A. Miller and E. Abrahams, Phys. Rev. **120**, 745 (1960).

[34] S. Chandrasekhar, Rev. Mod. Phys. **15**, 1 (1943).

[35] J. A. McInnes, P. N. Butcher and J. D. Clark, Phil. Mag. **B41**, 1 (1980).

[36] R. Landauer, "Electrical Conductivity in Inhomogeneous Media," in *Proceedings of the First Conference on the Electrical Transport and Optical Properties of Inhomogeneous Media*, edited by J. C. Garland and D. B. Tanner, (A.I.P. New York, 1978), p 2.

[37] S. Kirkpatrick, Rev. Mod. Phys. **45**, 574 (1973).

NONLINEAR EVOLUTION EQUATIONS WITH RAPIDLY OSCILLATING INITIAL DATA

D. McLaughlin, G. Papanicolaou
Courant Institute of Mathematical Sciences

O. Pironneau
University of Paris XI

Abstract

We give a brief description of how one can analyze the behavior of solutions of nonlinear equations when the initial data oscillate very rapidly.

* * *

Let $u(t,x)$ and $p(t,x)$ be the velocity and pressure, respectively, of a viscous incompressible fluid at $x \in \mathbb{R}^3$ and at time $t \geq 0$. The Navier-Stokes equations

$$(1) \qquad u_t + u \cdot \nabla u + \nabla p = \nu \Delta u, \qquad \nabla \cdot u = 0$$

are satisfied with ν the kinematic viscosity. Suppose that at time $t = 0$ the velocity field is made up of a mean part and a fluctuating part. How does such an initial flow evolve?

To be more precise we suppose that

$$(2) \qquad u(o,x) = V(x) + W(\tfrac{x}{\varepsilon})$$

where $V(x)$ is a given smooth vector function and $W(y)$ is a periodic or stationary random vector function with mean zero. The small parameter ε is the ratio of the scales of variation of the two parts of the flow when the scale of variation of V is one. The Reynolds number $Re = LU/\nu$, where L is a length scale and U a velocity scale, can be calculated with $L = 1$ or $L = \varepsilon$ i.e., relative to the mean flow or relative to the fluctuating flow. To make the problem physically interesting the Reynolds number relative to the fluctuating flow must not be small. Therefore, we let ν be proportional to ε ($\nu \to \varepsilon \nu$ in (1)). Now the question is: how does $u(t,x)$, $t > 0$, behave as ε tends to zero?

The answer depends in general on two additional facts. First on whether $W(y)$ is periodic or random, which means on whether the power spectrum of W has zero as an isolated element or not. This is important because in the periodic case there is a clean separation of scales between the two parts of the flow which does not exist otherwise.

Second the behavior as $\varepsilon \to 0$ depends on the nature of the solutions of

(3) $$v_t + v \cdot \nabla v + \nabla P = 0, \qquad \nabla \cdot v = 0,$$
$$v(o,x) = V(x)$$

which is the Euler equation for the mean flow. If v is smooth in some time interval we have one kind of behavior and if v has singularities another. Again, if the solution v is smooth there is only one scale of variation associated with it. If it has singularities there is scale mixing because v changes then rapidly and cannot be neatly distinguished from the fluctuating part of the flow.

In the simplest case: $v(t,x)$ smooth in $[o,T]$ and $W(y)$ periodic, we have the following result.

The solution $u^\varepsilon(t,x)$ of

(1') $$u_t^\varepsilon + u^\varepsilon \cdot \nabla u^\varepsilon + \nabla p^\varepsilon = \varepsilon \nu \Delta u, \qquad \nabla \cdot u^\varepsilon = 0$$
$$u^\varepsilon(o,x) = V(x) + W(\tfrac{x}{\varepsilon})$$

converges to the solution $v(t,x)$ of (3) exponentially fast as $\varepsilon \to 0$ and

(4) $$\lim_{\varepsilon \downarrow 0} \varepsilon \log |u^\varepsilon(t,x) - v(t,x)| = -S(t,x), \qquad 0 \le t \le T,$$

where $S(t,x)$ satisfies

(5) $$S_t + v \cdot \nabla S + \tfrac{\nu}{2}(\nabla S)^2 = \tfrac{\nu}{2}(\nabla \psi \cdot e)^2$$
$$S(o,x) = 0$$

(6) $$\psi_t + (v + \nu \nabla S) \cdot \nabla \psi = 0, \qquad \psi(o,x) = x$$

Here e is a unit vector in one of the coordinate directions and it is assumed that in the interval $(o,T]$, the positive function $S(t,x)$ takes its minimum value for this choice of e.

The way one obtains (4) is by looking for a solution of (1') in the form

(7) $$u^\varepsilon(t,x) = v(t,x) + \sum_p e^{-S^{(p)}(t,x)/\varepsilon} \left[W^{(p)} \left(\frac{\psi^{(p)}}{\varepsilon}, x, t \right) \right.$$
$$\left. + \varepsilon \hat{W}^{(p)} \left(\frac{\psi^{(p)}}{\varepsilon}, x, t \right) + \ldots \right]$$

with $\psi^{(p)}(t,x)$ and $S^{(p)}(t,x)$ suitably determined in the expansion process.

We see therefore that in the simplest case of clean separation of scales there is very little energy transfer between the mean and fluctuating flow as one would expect.

How does one analyze situations where substantial energy transfer exists? This is a difficult question that can so far be answered only for model problems such as Burgers' equation [1],[2].

One can obtain some understanding of energy transfer by assuming that a fluctuating external force acts on the fluid. The type of fluctuating force one assumes is of course determined by what kind of energy transfer mechanism one wants to excite. The notion of an external turbulent force is not a primary physical one. One visualizes the interaction of fluid masses with different characteristic length scales by means of forces acting on bigger scales to stir them up or on smaller scales to keep them going. Therefore probing a fluid with a turbulent force can be viewed as an intermediate step in a grand self-consistent scheme that explains energy transfer between fluid elements with different characteristic length scales. The mathematical implementation of such a scheme is extremely difficult and almost nothing is known about it.

The force we choose is as follows. The Navier-Stokes equations are

(8) $$u_t + u^\varepsilon \cdot \nabla u^\varepsilon = \nabla \cdot \tau + \frac{1}{\varepsilon} F\left(\frac{\theta^{-1}}{\varepsilon}, x, t \right) \cdot \nabla \theta^{-1}$$

$$\nabla \cdot u = 0$$

$$u(0,x) \text{ given,}$$

and the stress tensor τ is

(9) $$\tau_{ij} = -p^\varepsilon \delta_{ij} + \varepsilon \nu (u^\varepsilon_{i,j} + u^\varepsilon_{j,i}), \qquad i,j = 1,2,3 \;.$$

The functions $\theta(t,x)$ and $\theta^{-1}(t,x)$ are defined below.

We look now for a solution of (8) in the form

$$(9) \quad u_i(t,x) = v_i(t,x) + \theta_{i,K}(t,\theta^{-1}(t,x))\left[W_k\left(\frac{\theta^{-1}(t,x)}{\epsilon},x,t\right) + \epsilon\tilde{u}_k\left(\frac{\theta^{-1}(t,x)}{\epsilon},x,t\right)\right]$$

$$(10) \quad p^\epsilon(t,x) = \pi\left(\frac{\theta^{-1}(t,x)}{\epsilon},x,t\right) + P(t,x) + \epsilon\tilde{p}^\epsilon\left(\frac{\theta^{-1}(t,x)}{\epsilon},x,t\right).$$

We employ the summation convention and use the notation comma-subscript for derivative. Capital letter subscript means derivative with respect to the argument which is evaluated at θ^{-1}. When the expressions (9) and (10) are inserted in (8) and powers of ϵ are sorted out, we find that the fields W,π and v,P must satisfy the following equations.

$$(11) \quad K_{in}K_{ik}W_m W_{k,m} + \pi_{,n} = \nu K_{ik}K_{in}K^{-1}_{mj}K^{-1}_{\ell j}W_{k,j\ell} + F_n$$

$$W_{k,k} = 0,$$

$$\langle\pi\rangle = 0, \quad \langle F_n\rangle = 0, \quad \langle W_m\rangle = 0$$

$$(12) \quad v_t + v\cdot\nabla v = \nabla\cdot T, \quad \nabla\cdot v = 0, \quad v(0,x) = v_0(x),$$

$$(13) \quad T_{ij} = -P\delta_{ij} - \theta_{j,p}\theta_{i,k}\langle W_p(\cdot,\nabla\theta)W_k(\cdot,\nabla\theta)\rangle$$

$$(14) \quad \theta_t = v(t,\theta), \quad \theta(0,x) = x.$$

In (11) $W_m = W_m(y;K)$ where K is a 3×3 nonsingular matrix on which W depends parametrically and $F_n(y)$ is the nth component of the given external force assumed to be stationary random process with mean zero. Mean values or expectations are denoted by $\langle\ \rangle$. One can show that for each K nonsingular and $\nu > 0$ (11) has a unique solution in a suitable space which is a stationary random field, is divergence free and has mean zero. Once $W(y;K)$ has been obtained one calculates

$$(15) \quad C_{pk}(K) = \langle W_p(y;K)W_k(y;K)\rangle,$$

which is independent of y by stationarity, and then the effective stress tensor is given by (13) or

$$(16) \quad T_{ij} = -P\delta_{ij} - \theta_{j,p}\theta_{i,k}C_{pk}(\nabla\theta).$$

The mean velocity $v(t,x)$ satisfies (12) which is the Euler equation with a new component in the stress tensor due to the external force. The mapping $x \to \theta(t,x)$ of $\mathbb{R}^3$ to $\mathbb{R}^3$ is just the orbit of the fluid particle at time t as (14) shows.

We see now the effect of the special external random force on the flow as follows. From (8) the force field is assumed to be transported along the characteristics of the mean flow. The fluctuation or microstructure field W is determined locally by its own steady Navier-Stokes equation (11) relative to a coordinate system adapted to the characteristics of the mean flow (parametric dependence on K). The microstructure flow W is affected by the mean flow through the slow changing of K ($K = \nabla\theta(t,x)$). The mean or macroscopic flow $v(t,x)$ satisfies the inviscid Euler equation (12) with effective stress tensor (13) or (16). The microstructure affects the macrostructure through the "Reynolds stress" term $\nabla\theta\nabla\theta \, C(\nabla\theta)$ in (16).

Depending on the type of force F one has, the new term in the effective stress in (16) may correspond to energy transfer to the mean flow or energy transfer out of it. It is very hard to say which forces produce a dissipative and which a pumping effect. Our main result is that if (12),(13),(14) have a smooth solution in some finite time interval then (11) is well defined for F small enough and the whole solution u^ε of (8) is well represented by (9),(10).

There are two issues that must be settled now in connection with the external force problem. The first one is the characterization of forces that lead to a dissipative mean stress tensor as mentioned above. The second and harder problem is the determination of a self-consistent scheme for energy transfer in a realistic situation in which each step in some iteration process is a problem of the form (8) with the force F determined through the flow at a previous step. This sounds like some kind of renormalization group analysis [3] but concrete results seem hard to get at present.

A number of other problems, mostly model problems that give rise to Reynolds stress terms, have been worked out [4] partly in connection with the theory of turbulent diffusion. Other nonlinear equations with rapidly oscillating initial data are considered in [5].

Note: This work was supported by the Air Force Office of Scientific Research, Grant AFOSR-80-0228.

REFERENCES

[1] J.M. Burgers, The Nonlinear Diffusion Equation, D. Reidel Publishing Co., 1974.

[2] S. Kida, Asymptotic properties of Burgers turbulence, J. Fluid Mech. (1979) 93, 2, pp. 337-377.

[3] J.-D. Fournier, Quelque methodes systematiques d'approximation en turbulence homogene, These, Université de Nice, 1977.

[4] D. McLaughlin, G. Papanicolaou and O. Pironneau, Selfconsistent advection of microstructure in viscous fluids. (To appear.)

[5] D. McLaughlin, G. Papanicolaou and L. Tartar, Weak limits of conservation laws with oscillating data, SIAM J. Appl. Math. (To appear.)

A COMPARISON OF TWO METHODS FOR DERIVING BOUNDS ON
THE EFFECTIVE CONDUCTIVITY OF COMPOSITES

G. W. Milton
Laboratory of Atomic and Solid State Physics and
Materials Science Center, Clark Hall,
Cornell University, Ithaca, NY, 14853, U.S.A.
and
R. C. McPhedran
Department of Theoretical Physics,
The University of Sydney, Sydney, N.S.W., 2006, Australia

Abstract

An infinite set of bounds on the effective conductivity σ_e of a two-component composite has previously been derived using the analytic properties of σ_e as a function of σ_1 and σ_2 (the conductivities of the components). We show that this same set of bounds can alternatively be derived from variational principles. The bounds incorporate information about the microstructure of the composite in addition to the volume fractions of the components.

I. Introduction

There are two dominant methods for deriving rigorous bounds on the effective conductivity σ_e of a two-component composite. The first method, largely developed by Hashin and Shtrikman,[1] Brown[2] and Beran,[3] involves the use of variational principles. A second method, pioneered by Bergman,[4] makes use of the analytic properties of σ_e as a function of the conductivities σ_1 and σ_2 of the components.

Using the analytic method, an infinite set of bounds on σ_e has been derived by Milton.[5] The appropriate pair of bounds (one an upper bound and the other a lower bound) depend on the known structure and properties of the composite. Each bound-pair is characterized by an integer pair (L,M): L signifies that the bound-pair incorporates L structural parameters; M represents the number of known values of σ_e (for various values of σ_1 and σ_2) which are incorporated in the bounds. The set of bounds (1,0), (2,0) and (2,1) were originally derived using the analytic method by Bergman.[4]

In this paper we show that the entire set of bounds (L,0) where L is arbitrary can also be derived using variational principles. First we briefly describe these bounds.

II. Description of the (L,O) bounds

It is convenient to introduce a new variable

$$\rho = (\sigma_1 - \sigma_2)/(\sigma_1 + \sigma_2), \tag{1}$$

and to assume, without loss of generality that

$$\sigma_1 \geq \sigma_2 = 1. \tag{2}$$

The Lth order bound-pair incorporates the values of the structural parameters

$$Q_n = \left.\frac{d^n \sigma_e(e)}{d\rho^n}\right|_{\rho = 0}, \tag{3}$$

for $n = 1$ to $n = L$. It follows from the work of Brown[6] that for any three-dimensional isotropic composite material,

$$Q_1 = 2f_1 \tag{4}$$

$$Q_2 = 4f_1(1 + 2f_1)/3 \tag{5}$$

$$Q_3 = 16f_1^2 - 4f_1 + 16f_1 f_2 (2\zeta_1 + f_2)/3 \tag{6}$$

where $f_1 = 1 - f_2$ is the volume fraction occupied by component 1 and ζ_1 is a geometric parameter which can be written in terms of the three-point correlation function (not usually known) and can have any value in the range [0,1]. Figure 1 shows how ζ_1 depends on f_1 for periodic systems, in particular the simple cubic (SC), body centered cubic (BCC) and face centered cubic (FCC) lattices of spheres of uniform radius. This same parameter ζ_1 is also incorporated in bounds on the effective elastic moduli of two-component composites.

The values of the Q_n for higher values of n can in principle be determined from the n-point correlation function.[6] However, if $n \geq 2$, the n-point correlation function contains a vast amount of statistical information, not all of which is relevant to the determination of $\sigma_e(\sigma_1, \sigma_2)$. Indeed, composites with different geometrical structures may have the same characteristic function $\sigma_e(\sigma_1, \sigma_2)$. It is appropriate to think of the set of parameters Q_n as representing the relevant details of the microstructure of the composite.

Fig. 1 The structural parameter ζ_1 as a function of the volume fraction f_1 of the inclusions for periodic lattices of spheres of uniform size. The curves from left to right correspond to SC, BCC and FCC lattices and each arrow denotes the volume fraction when the spheres touch. The graphs are plotted from the tabulated data of McPhedran and Milton.[7]

Numerical studies by McPhedran and Milton[7] have shown that provided σ_1/σ_2 is not zero or not infinite, the bounds converge rapidly to the exact value of σ_e as progressively more information is supplied, i.e., as L is increased. These studies have been carried out again for periodic systems, for SC, FCC and BCC lattices of spheres of uniform radius and for square and hexagonal arrays of cylinders. The convergence properties persist even for ratios of σ_1/σ_2 as large as 10,000:1 (Figs. 2,3). As yet we do not know of any analytic proof that the bounds always converge to the precise value of σ_e (provided $\sigma_1/\sigma_2 \neq 0$ or ∞), but it is conjectured that this is the case.

The prescription for caluclating the bounds depends on whether L is even or odd. In either situation the bounds can be expressed in terms of the following function:

$$g(\rho) = (1 + \sum_{h=1}^{N} \alpha_h \rho^h)/(1 + \sum_{h=1}^{N} \beta_h \rho^h). \qquad (7)$$

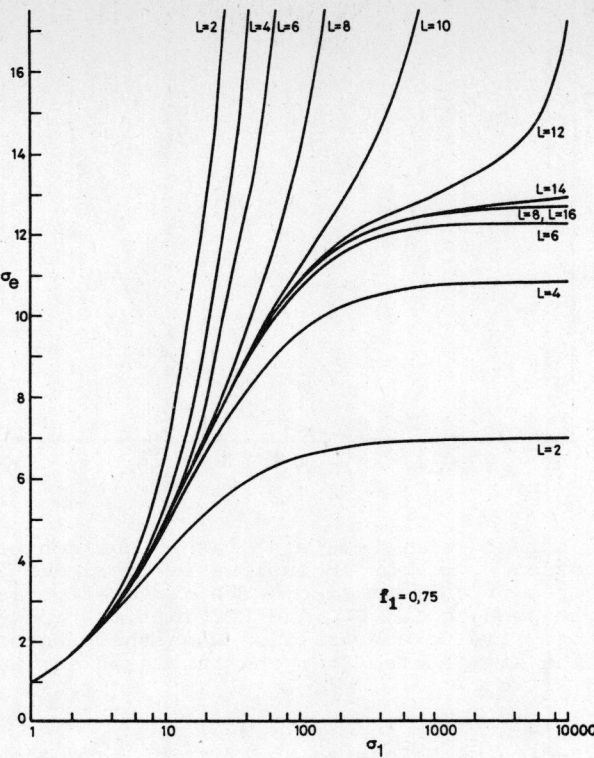

Fig. 2 The upper and lower Lth order bounds on the transverse conductivity of a square array of circular cylinders of uniform diameter, conductivity σ_1, in a matrix of unit conductivity. The cylinders occupy 75% of the composite and so are close to touching. All the bounds, for even L, are attainable by geometries consisting of multicoated cylinders.[5] The graph is taken from McPhedran and Milton.[7]

Here the 2N real coefficients α_h and β_h are such that

$$\left.\frac{d^n g}{d\rho^n}\right|_{\rho=0} = Q_n \quad (n=1,2\ldots,L). \tag{8}$$

If L is even and $2N = L$, then the α_h and β_h are determined by the 2N constraints (8). In this case $g(\rho)$ is a <u>lower bound</u> on σ_e. Alternatively if $2N = L + 2$ and we impose the additional constraints

$$g(1) = \infty, \tag{9}$$
$$g(-1) = 0, \tag{10}$$

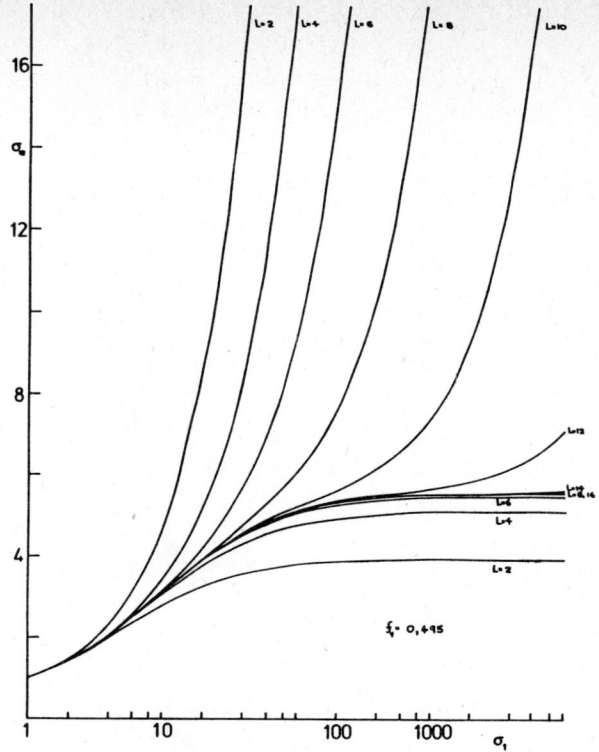

Fig. 3 As for Fig. 2, but for a cubic array of spheres occupying 49.5% of the composite. It is not known if these bounds are attainable. The graph is taken from McPhedran and Milton.[7]

then $g(\rho)$, which is again uniquely determined, is an <u>upper bound</u> on σ_e.

If L is odd and $2N = L + 1$, then $g(\rho)$ is an <u>upper or lower bound</u> on σ_e, depending on whether we impose the addition constraint (9) or the constraint (10).

As an example of the use of these bounds, let us take $L = 1$. We find that

$$\frac{1 + \rho}{1 - (f_1 - f_2)\rho} \leq \sigma_e \leq \frac{1 + (f_1 - f_2)\rho}{1 - \rho} , \qquad (11)$$

or equivalently

$$[f_1/\sigma_1 + f_2]^{-1} \leq \sigma_e \leq f_1\sigma_1 + f_2 . \qquad (12)$$

The bounds (12) are precisely the same as those derived by Wiener.[8] Similarly the L = 2, L = 3, L = 4, and L = 5 bounds correspond exactly with the bounds derived by Hashin and Shtrikman,[1] Beran,[3] Phan-Thien and Milton,[9] and Elsayed.[10] These bounds were derived using variational principles.

We will now show that the entire set of bounds, for arbitrary L, can in fact be derived from variational principles.

III. Derivation of bounds using classical variational principles

According to one of the classical variational principles, if we choose any trial electric field $\underline{E}^T$ such that

$$\langle \underline{E}^T \rangle = \hat{\underline{x}}, \tag{13}$$

$$\underline{\nabla} \times \underline{E}^T = 0, \tag{14}$$

where $\hat{\underline{x}}$ is a unit vector in the x direction, and the brackets $\langle \ \rangle$ denote a volume average, then the quantity

$$\sigma_{eU} = \langle \sigma \underline{E}^T \underline{E}^T \rangle, \tag{15}$$

is an upper bound on σ_e. The proof is straightforward. If $\underline{E}$ denotes the actual electric field (with $\langle \underline{E} \rangle = \hat{\underline{x}}$) and $\underline{E}^* = \underline{E}^T - \underline{E}$, then

$$\sigma_{eU} = \langle \sigma \underline{E}\,\underline{E} \rangle - 2\langle \sigma \underline{E}^* \underline{E} \rangle + \langle \sigma \underline{E}^* \underline{E}^* \rangle. \tag{16}$$

By applying Green's theorem we find

$$\langle \sigma \underline{E}\,\underline{E} \rangle = \sigma_e, \tag{17}$$

$$\langle \sigma \underline{E}^* \underline{E} \rangle = 0, \tag{18}$$

and hence

$$\sigma_{eU} = \sigma_e + \langle \sigma \underline{E}^* \underline{E}^* \rangle, \tag{19}$$

which is clearly greater than σ_e.

The actual electric field $\underline{E}$ will be an analytic function of ρ, and so if ρ is small we can write

$$\underline{E}(\rho) = \sum_{n=0}^{\infty} \rho^n \underline{E}_n \tag{20}$$

where the fields E_n are independent of ρ, and $E_0 = \hat{x}$. Following the method outlined by Beran,[3] we take as our trial field

$$E^T = \sum_{n=0}^{M} a_n E_n \qquad (21)$$

where, from (13), we require that $a_0 = 1$. The other coefficients a_n will be chosen to minimize σ_{eU}. Observe that

$$\sigma = 1 + \Omega(\sigma_1 - 1) = [2\rho\Omega + (1 - \rho)]/(1 - \rho) \qquad (22)$$

where $\Omega(z)$ is one if z is in component 1 and zero otherwise.

Substituting (21) in (15) we find

$$\sigma_{eU} = [1 + \rho(f_1 - f_2) + H_{ij}a_i a_j + 2L_i a_i]/(1 - \rho), \qquad (23)$$

where the summation over i and j extends from 1 to M and

$$H_{ij} = 2\rho <\Omega\, E_i E_j> + (1 - \rho) <E_i E_j> \qquad (24)$$

$$L_i = 2\rho <\Omega E_i \hat{x}> + (1 - \rho) <E_i \hat{x}> \qquad (25)$$

The best possible choice of $a = (a_1, a_2 \ldots a_m)$ is found by setting $\partial \sigma_{eU}/\partial a = 0$ giving

$$a = -H^{-1} L \qquad (26)$$

and substituting this in (23),

$$\sigma_{eU} = [1 + \rho(f_1 - f_2) - L^T H^{-1} L]/(1 - \rho), \qquad (27)$$

where L^T is the transpose of L.

Note that all the elements of the matrix H and the column vector L are linear functions of ρ. It follows that $\sigma_{eU}(\rho)$, given by (27), can be expressed in the form (7) where $N = M + 1$. Also, from (27),

$$\sigma_{eU}(1) = \infty. \qquad (28)$$

If instead of choosing the optimum values, we had taken $a_h = \rho^h$ then E^* would be of order ρ^{M+1} and consequently from (19) $\sigma_{eU} - \sigma_e$ would be of order $\rho^{2(M+1)}$. The bounds (27) being more restrictive must similarly coincide with the exact value of σ_e up to terms of order

ρ^{2M+1}. Consequently

$$\left.\frac{d^n \sigma_{eU}}{d\rho^n}\right|_{\rho=0} = Q_n \quad (n=1,2,\ldots,2M+1). \tag{29}$$

Since $\sigma_{eU}(\rho)$ can be expressed in the form (7) where $N = M + 1$, the $2(M+1)$ constraints (28) and (29) are sufficient to determine the function $\sigma_{eU}(\rho)$. Thus if we know Q_n, for $n = 1$ up to $n = 2M + 1$, we can calculate the bound σ_{eU}. We do not need to evaluate the quantities $\langle \Omega E_i E_j \rangle$, $\langle E_i E_j \rangle$, $\langle \Omega E_i \hat{x} \rangle$ and $\langle E_i \hat{x} \rangle$. The bound $\sigma_{eU}(\rho)$ is identical to the Lth order upper bound where $L = 2M + 1$.

The other classical variational principle states that if we choose any trial current field $\underline{J}^T$ such that

$$\langle \underline{J}^T \rangle = \hat{\underline{x}}, \tag{30}$$

$$\underline{\nabla} \cdot \underline{J}^T = 0, \tag{31}$$

then

$$\sigma_{eL} = 1/\langle \underline{J}^T \underline{J}^T/\sigma \rangle \tag{32}$$

is a lower bound on σ_e. Following Beran[3] we express the true current field $\underline{J}$ as a power series in ρ

$$\underline{J} = \sum_{n=0}^{\infty} \underline{J}_n \rho^n, \tag{33}$$

where $\underline{J}_0 = \hat{\underline{x}}$. We take as our trial field

$$\underline{J}^T = \sum_{n=0}^{M} b_n \underline{J}_n, \tag{34}$$

where from (30) we require $b_0 = 1$. Provided $\underline{b} = (b_1, b_2, \ldots, b_M)$ is chosen to maximize σ_{eL} we find that $\sigma_{eL}(\rho)$ coincides with the Lth order lower bound, where $L = 2M + 1$.

IV. Bounds derived from the Hashin-Shtrikman Variational Principle

According to the Hashin-Shtrikman variational principle,[1] if we take any trial "polarization field" $\underline{P}^T$ and let $\underline{A}^T$ denote the associated field which is determined by the constraints

$$\sigma_0 \nabla \cdot \underline{A}^T + \nabla \cdot \underline{P}^T = 0, \qquad (35)$$

$$\nabla \times \underline{A}^T = 0, \qquad (36)$$

$$\langle \underline{A}^T \rangle = 0, \qquad (37)$$

where σ_0 is a constant, then

$$\sigma_{eb} = \langle \sigma_0 - \frac{\underline{P}^T \cdot \underline{P}^T}{(\sigma - \sigma_0)} + 2\underline{P}^T \cdot \hat{\underline{x}} + \underline{P}^T \cdot \underline{A}^T \rangle \qquad (38)$$

is an upper bound on σ_e when $\sigma_0 \geq \sigma_1$ and a lower bound when $\sigma_0 \leq \sigma_2(=1)$. Let us take $\sigma_0 = \sigma_2 = 1$ and choose as our trial field

$$\underline{P}^T = (\sigma - \sigma_0)\underline{E}^T = (\sigma_1 - 1)\left(\sum_{n=0}^{M} a_n \Omega \underline{E}_n\right), \qquad (39)$$

in which case

$$\underline{A}^T = \tfrac{1}{2}(\sigma_1 - 1)\left[a_0 \hat{\underline{x}} + \sum_{n=0}^{M} a_n(\underline{E}_{n+1} - \underline{E}_n)\right]. \qquad (40)$$

Substituting (39) and (40) in (38) and selecting the optimum value of $\underline{a} = (a_0, a_1 \ldots, a_m)$ we find that

$$\sigma_{eb} = 1 - 2\rho(\underline{G}^T \underline{F}^{-1} \underline{G}), \qquad (41)$$

where

$$G_i = \langle \Omega \underline{E}_i \hat{\underline{x}} \rangle, \qquad (42)$$

$$F_{ij} = \rho(\delta_{i0} \langle \Omega \underline{E}_j \hat{\underline{x}} \rangle + \langle \Omega \underline{E}_i \underline{E}_{j+1} \rangle) - \langle \Omega \underline{E}_i \underline{E}_j \rangle. \qquad (43)$$

Since i and j range from 0 to M and all the matrix elements of F_{ij} are linear in ρ, it follows that σ_{eb} can be expressed in the form (7) where $N = M + 1$.

Now denote $\underline{P}^* = \underline{P}^T - \underline{P}$ and $\underline{A}^* = \underline{A}^T - \underline{A}$ where $\underline{P} = (\sigma - \sigma_0)\underline{E}$ and $\underline{A} = \underline{E} - \hat{\underline{x}}$. As Hashin and Shtrikman[1] prove

$$\sigma_{eb} - \sigma_e = \langle \underline{A}^* \cdot \underline{P}^* - \frac{(\underline{P}^* \cdot \underline{P}^*)}{(\sigma - \sigma_0)} \rangle. \qquad (44)$$

If we had taken $a_n = \rho^n$ then $\underline{P}^*$ would be of order ρ^{M+2}, $\underline{A}^*$ of order ρ^{M+1} and $\sigma_{eb} - \sigma_e$ of order ρ^{2M+3}. The bound (41) being more restrictive must also coincide with σ_e up to terms of order ρ^{2M+2}.

Thus

$$\left.\frac{d^n \sigma_{eb}}{d\rho^n}\right|_{\rho=0} = Q_n \quad (n=1,2,\ldots,2M+2). \tag{45}$$

The $2(M+1)$ constraints (45) are sufficient to determine the function σ_{eb}, since it can be expressed in the form (7) where $N = M + 1$.

We have established that the bound σ_{eb} coincides with the Lth order lower bound, where $L = 2(M+1)$. Similarly, if $\sigma_0 = \sigma_1$, the bound σ_{eb} is identical with the Lth order upper bound, where $L = 2(M+1)$.

V. Conclusion

In summary, if L is odd then the Lth order upper and lower bounds can be derived using the classical variational principles. If L is even, then the Lth order upper and lower bounds can be derived using the Hashin-Shtrikman variational principles. If L is either even or odd then the bounds can be derived using the analytic method.

Thus the analytic method and the variational method are equally powerful when applied to the problem of finding bounds on σ_e, for real σ_1 and σ_2, given the values of Q_n for $n = 1$ up to $n = L$. However, each method has its own advantages: the analytic method seems most suitable for deriving bounds on σ_e when σ_1 and σ_2 are complex;[11,12] the variational method appears most appropriate for deriving bounds on σ_e for multi-component materials.

Achnowledgments

This work was supported in part by the Department of Energy through a grant (Grant #XH-9-8158-1) from the Solar Energy Research Institute. It was undertaken while one of the authors (G.W.M.) was a recipient of a Sydney University Traveling Scholarship.

References

1. Z. Hashin and S. Shtrikman, J. Appl. Phys. 33, 3125 (1962).
2. W. F. Brown, Trans. Soc. Rheol. 9, 357 (1965).
3. M. Beran, Nuovo Cimento 38, 771 (1965).
4. D. J. Bergman, Phys. Rep. 43, 377 (1978).
5. G. W. Milton, J. Appl. Phys., in press.
6. W. F. Brown, J. Chem. Phys. 23, 1514 (1955).
7. R. C. McPhedran and G. W. Milton, submitted to Appl. Phys.

8. O. Wiener, Abh. Math. Phys. Kl. Königl, Saechs. Ges. 32, 509 (1912).

9. N. Phan-Thien and G. W. Milton, submitted to Proc. Roy. Soc. Lond.

10. M. A. Elsayed, J. Math. Phys. 15, 2001 (1974).

11. D. J. Bergman, Phys. Rev. Lett. 44, 1285 (1980).

12. G. W. Milton, Appl. Phys. Lett. 37, 300 (1980).

FLUCTUATION CORRECTIONS TO THE MEAN FIELD
DESCRIPTION OF A NONUNIFORM FLUID

J. K. Percus*
Courant Institute of Mathematical Sciences
and Department of Physics
New York University
New York, N. Y. 10012

The interaction of a fluid of particles is separated into a core-core repulsion and a communally generated attractive field. Existence of a self-maintained density profile between liquid and gas phases is investigated in the Van der Waals model, and then from a microscopic point of view under the assumptions of local homogeneity and fluctuationless mean field. The particle degrees of freedom are rigorously eliminated, and the pure field system analyzed first under the previous assumptions and then in the presence of leading order field fluctuations. Divergent broadening of the density profile in two or three dimensional space is found, and related to the capillary wave picture.

1. Introduction

I would like to discuss the propagation of a very special field through a special disordered medium. The medium can be imagined as a gas of (perhaps soft) microscopic billiard balls. Except to the expert, it is not a very interesting fluid. As a gas, it occupies an arbitrarily large volume under sufficiently low pressure. Increasing its density requires proportionally increasing the pressure, at first, and then increasing it more and more rapidly as the available volume gets used up. But suppose now that each billiard ball is also the source of an attractive force field, so that each ball or core is immersed in the communal force field of the others. Then as the density is increased at fixed temperature, each core finds itself eventually sitting in an increasingly deep, somewhat erratic, potential well from which it escapes only occasionally -- into another well -- if its thermal energy is not too large. For this trapped or liquid state, the system pressure can be quite low, and so there is the possibility of a fluid at a given temperature and pressure possessing either a low density disordered gas phase, a high density more or less ordered liquid phase, or both, in spatially separate enclaves.

The spatial transition between two phases, as a self-supported rapidly changing density, must have some unique properties, and it is these that we want to focus on. The possibility of maintaining a density gradient in the absence of external forces can be assessed quite easily in a preliminary way. A unit volume of billiard balls at $\underset{\sim}{r}$ will be subject to an average force $-\nabla P_0(n(\underset{\sim}{r}))$, where $P_0(n)$ is the equation of state and $n(\underset{\sim}{r})$ the spatially varying density. The average potential, due to the attractive force field, felt by a particle in this volume is a linear functional of the density, and hence (if intrinsically isotropic) can be expanded as $-an(\underset{\sim}{r}) - b\nabla^2 n(\underset{\sim}{r}) + \ldots$. Thus, the average forces will be balanced if $-\nabla P_0(n(\underset{\sim}{r})) + n(\underset{\sim}{r})\nabla(an(\underset{\sim}{r}) + b\nabla^2 n(\underset{\sim}{r})) = 0$, which we shall rewrite as [1]

$$b\nabla^2 n(\underset{\sim}{r}) + an(\underset{\sim}{r}) - \mu_0(n(\underset{\sim}{r})) = -\mu \qquad (1.1)$$

where $\mu_0'(n) = P_0'(n)/n$.

μ is the constant of integration, and $\mu_0(n)$ -- called the chemical potential -- starts logarithmically with n and eventually increases in the same fashion as $P_0(n)$.

Eq. (1.1) is consistent with two different uniform densities n_1 (liquid) and n_2 (gas) only if

$$\mu(n_1) = \mu = \mu(n_2) \qquad (1.2)$$

where $\mu(n) \equiv \mu_0(n) - an$.

If a system is to have simultaneously $n(\underset{\sim}{r}) = n_1$, say as $z \to \infty$, and $n(\underset{\sim}{r}) = n_2$, say as $z \to -\infty$, (1.2) is not sufficient. Take $\partial/\partial z$ of (1.1) and multiply by $n(\underset{\sim}{r})$:

$$bn(\underset{\sim}{r})\nabla^2 \frac{\partial}{\partial z} n(\underset{\sim}{r}) = \frac{\partial}{\partial z} P(n(\underset{\sim}{r})) \qquad (1.3)$$

where $P(n) \equiv P_0(n) - \frac{1}{2} an^2$.

Since $n\nabla^2 \partial n/\partial z = \frac{1}{2} \nabla \cdot (n\nabla \partial n/\partial z - \nabla n \partial n/\partial z) + \frac{1}{2}\partial(n\nabla^2 n)/\partial z$, integration over a large rectangular parallelopiped yields as well the requirement

$$P(n_1) = P(n_2) . \qquad (1.4)$$

Indeed $\mu'(n) = P'(n)/n$, and (1.2, 1.4) are readily identified with equality of pressure and chemical potential for bulk gas and liquid.

If (1.1, 1.2, 1.4) are satisfied, a spatial two-phase separation does occur, whose form we shall examine later. Our objective now will be two-fold: 1) to put the mean field approximation implicit in (1.1) on a somewhat firmer basis, and 2) to examine the nature of the field fluctuations which have been neglected in (1.1), and their effect on observable quantities.

2. Mean Field Approximation

The conceptual division referred to above -- particle cores immersed in a self-generated force field -- can be developed, although not uniquely, into an effective computational tool. One possibility is to describe the system jointly by the locations $\{r_i\}$ of its particles -- coupled by some "billiard-ball" interaction potential $\phi_0(r_i - r_j)$ -- and the potential field $v(r)$ resulting from the additional attractive forces. Thus for the dynamics of the particles,

$$m\ddot{r}_i = -\sum_j \nabla \phi_0(r_i - r_j) - \nabla v(r_i) , \qquad (2.1)$$

with the convention $\nabla \phi_0(0) = 0$. If the attractive potential giving rise to the field is written as $-\phi_1(r_i - r_j)$ for a given particle pair, then of course

$$v(r) = -\sum_j \phi_1(r - r_j) , \qquad (2.2)$$

where now $\nabla \phi_1(0) = 0$. Suppose that $\phi_1(r - r')$, as the kernel of an integral operator, is invertible. Then we can rewrite (2.2) -- including a propagation delay mechanism which vanishes as the field mass density $M \to 0$ -- as

$$M\ddot{v}(r) = -\int \phi_1^{-1}(r - r') v(r') dr' - \sum_j \delta(r - r_j) . \qquad (2.3)$$

The reason for doing this is that then both (2.1) and (2.3) are consequences of the equations of motion for the Hamiltonian [2] (p_i = particle momentum, $\pi(r)$ = field momentum)

$$H = \sum_i p_i^2/2m + \frac{1}{2} \sum_{i,j} \phi_0(r_i - r_j) + \sum_i v(r_i)$$
$$+ \frac{1}{2} \iint \phi_1^{-1}(r - r') v(r) v(r') dr dr' + \int \pi(r)^2/2M dr , \qquad (2.4)$$

as is readily verified. Furthermore, if the total energy is rewritten by substituting v from (2.3) in (2.4), there results

$$E = \sum_i p_i^2/2m + \int \pi(\underline{r})^2/2M \, d\underline{r} + \frac{1}{2} \iint \dot{\pi}(\underline{r}) \phi_1(\underline{r}-\underline{r}') \dot{\pi}(\underline{r}') d\underline{r} d\underline{r}'$$
$$+ \frac{1}{2} \sum_{i,j} \left(\phi_0(\underline{r}_i - \underline{r}_j) - \phi_1(\underline{r}_i - \underline{r}_j) \right) , \quad (2.5)$$

the coordinate part of which is the full particle interaction energy.

In this paper, we will restrict attention to time-average equilibrium properties. Introducing the microscopic particle density

$$\rho(\underline{r}) \equiv \Sigma \delta(\underline{r} - \underline{r}_i) , \quad (2.6)$$

the average particle density and particle pair density defined by

$$n(\underline{r}) = \langle \rho(\underline{r}) \rangle$$
$$n_2(\underline{r},\underline{r}') = \langle \sum_{i \neq j} \delta(\underline{r}-\underline{r}_i) \delta(\underline{r}'-\underline{r}_j) \rangle \quad (2.7)$$
$$= \langle \rho(\underline{r}) \rho(\underline{r}') - \rho(\underline{r}) \delta(\underline{r}-\underline{r}') \rangle$$

will be of principal interest, as well as the average field $\bar{v}(\underline{r}) \equiv \langle v(\underline{r}) \rangle$. For a <u>uniform</u> simple fluid of density n, the pair density is translation-invariant and isotropic, so that one has

$$n(\underline{r}) = n$$
$$n_2(\underline{r},\underline{r}') = n^2 g(|\underline{r}-\underline{r}'|) , \quad (2.8)$$

where g is the radial distribution function.

On averaging (2.3), we have to start with the exact

$$\int \phi_1^{-1}(\underline{r}-\underline{r}') \bar{v}(\underline{r}') d\underline{r}' = -n(\underline{r}) . \quad (2.9)$$

The approximations take place in (2.1), which we first multiply by $\delta(\underline{r}-\underline{r}_i)$ and average, obtaining (by use of $\langle \dot{A}B \rangle = -\langle A\dot{B} \rangle$)

$$m \sum_i \langle \dot{\underline{r}}_i \dot{\underline{r}}_i \cdot \nabla \delta(\underline{r}-\underline{r}_i) \rangle = -\int \nabla \phi_0(\underline{r}-\underline{r}') \langle \rho(\underline{r}) \rho(\underline{r}') \rangle d\underline{r}'$$
$$- \langle \nabla v(\underline{r}) \rho(\underline{r}) \rangle . \quad (2.10)$$

Since, at reciprocal temperature β, $\langle \dot{r}_{i\mu} \dot{r}_{i\nu} f(\underline{r}_i) \rangle = (\delta_{\mu\nu}/m\beta) \langle f(\underline{r}_i) \rangle$, this becomes, after a minor transformation,

$$\langle \nabla \rho(\underline{r}) \rangle / \beta - \frac{1}{2} \int \nabla \phi_0(\underline{r}') \langle \rho(\underline{r}+\underline{r}') \rho(\underline{r}) - \rho(\underline{r}) \rho(\underline{r}-\underline{r}') \rangle d\underline{r}'$$
$$= -\langle \nabla v(\underline{r}) \rho(\underline{r}) \rangle . \quad (2.11)$$

Two approximations are now required. The first one, the local thermodynamics approximation, has only to do with the properties of the underlying core fluid under the assumption that the density is slowly varying in space on the scale of the range of particle-particle correlations. It avers that under these circumstances, the pair density -- as well as other intensive quantities -- can be taken as that of the uniform system at the local density. The assumed slow variation thus means that (2.11) can be written as

$$-\langle v(\underline{r})\rho(\underline{r})\rangle = \nabla n(\underline{r})/\beta - \frac{1}{2}\nabla \cdot \int \underline{r}' \nabla \phi_0(\underline{r}') \langle \rho(\underline{r}+\underline{r}')(\rho(\underline{r})-\delta(\underline{r}'))\rangle d\underline{r}' \quad (2.12)$$

$$= \nabla[n(\underline{r})/\beta - \frac{1}{6} n(\underline{r})^2 \int \underline{r}' \cdot \nabla \phi_0(\underline{r}') g_0(\underline{r}';n(\underline{r})) d\underline{r}'] ,$$

but the bracketed expression is just the well-known virial form for the pressure, and so

$$\nabla P_0(n(\underline{r})) = -\langle \nabla v(\underline{r})\rho(\underline{r})\rangle . \quad (2.13)$$

The second approximation, the mean field approximation, is that the variations of the field $v(\underline{r})$ are negligible (or at least uncorrelated with $\rho(\underline{r})$), so that $v(\underline{r})$ can be replaced by its mean in (2.13):

$$\nabla P_0(n(\underline{r})) = -n(\underline{r}) \nabla \bar{v}(\underline{r}) . \quad (2.14)$$

From (1.1), this is equivalent to

$$\mu_0(n(\underline{r})) + \bar{v}(\underline{r}) = \mu \quad (2.15)$$

for a suitable constant μ. Hence if the inverse of $\mu_0(n)$ is designated by $n_0(\mu)$, (2.9) and (2.15) combine to give

$$\int \phi_1^{-1}(\underline{r}-\underline{r}')(\mu_0(\underline{r}') - \mu) d\underline{r}' = n_0(\mu_0(\underline{r})) \quad (2.16)$$

which can be solved at leisure. A model ϕ_1 makes the job easier and more explicit, and we shall choose the "Yukawa potential"

$$\phi_1(\underline{r}-\underline{r}') = K e^{-\gamma|\underline{r}-\underline{r}'|} / |\underline{r}-\underline{r}'| \quad (2.17)$$

with the convenient property that the inverse operator is simply

$$\phi_1^{-1} = \frac{1}{4\pi K}(\lambda^2 - \nabla^2) . \quad (2.18)$$

Eq. (2.16) now becomes

$$\frac{1}{4\pi K} \nabla^2 \mu_0 + n_0(\mu_0) - \frac{\lambda^2}{4\pi K}(\mu_0 - \mu) = 0 \ . \tag{2.19}$$

Let us look at the possibility of a planar interface. Thus (2.19) has only a z-dependence, and it can be reduced in differential order by multiplying by $\mu_0'(z)$ and integrating. This yields

$$\frac{1}{8\pi K} \mu_0'(z)^2 + P_0(\mu_0(z)) - \frac{\lambda^2}{8\pi K}(\mu_0(z)-\mu)^2 = P \tag{2.20}$$

for suitable constant P. Of course, we have simply found the "energy integral" for (2.19), and the resulting profile is determined by the "potential"

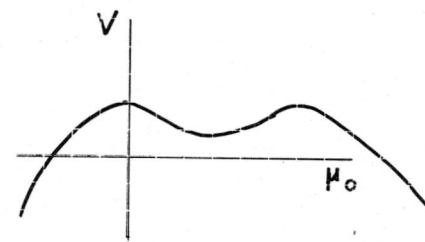

$$V(\mu_0) = P_0(\mu_0) - \frac{\lambda^2}{8\pi K}(\mu_0-\mu)^2 \tag{2.21}$$

In particular, asymptotically constant values of $\mu_0(\infty) = \mu_1$, $\mu_0(-\infty) = \mu_2$ are obtained providing that $V(\mu_1) = V(\mu_2)$ and $V'(\mu_1) = 0 = V'(\mu_2)$, which we write as

$$P_0(\mu_1) - \frac{\lambda^2}{8\pi K}(\mu_1-\mu)^2 = P = P_0(\mu_2) - \frac{\lambda^2}{8\pi K}(\mu_2-\mu)^2$$
$$\mu_1 - \frac{4\pi K}{\lambda^2} n_0(\mu_1) = \mu = \mu_2 - \frac{4\pi K}{\lambda^2} n_0(\mu_2) \ . \tag{2.22}$$

P and μ can be identified with the common pressure and chemical potential of the system with full interaction $\phi_0 - \phi_1$.

A prototypical example due to Van der Waals [1] (but in the context of (1.1), which is equivalent to (2.16) when $\phi_1 = a + b\nabla^2$) is that in which $P = \mu = 0$, $\mu_2 = n_0(\mu_2) = 0$, and $P_0(\mu_0) = (\lambda/\mu_1)^2 \mu_0^3 (2\mu_1 - \mu_0) / 8\pi K$, so that

$$V(\mu_0) = \frac{-1}{8\pi K} \left(\frac{\lambda}{\mu_1}\right)^2 \mu_0^2 (\mu_1 - \mu_0)^2 \ . \tag{2.23}$$

From (2.20), then, $\lambda \mu_1 dz/d\mu_0 = (1/\mu_0 + 1/(\mu_1 - \mu_0))$, producing the profile

$$\mu_0(z) = \frac{\mu_1}{1+e^{-\lambda z}} = \frac{\mu_1}{2}(1 + \tanh \tfrac{1}{2}\lambda z) \ . \tag{2.24}$$

As for the density, (2.16) now reads $n = (\lambda^2 - d^2/dz^2)\mu_0/4\pi K$, or

$$n(z) = \frac{n_1}{1+e^{-\lambda z}} \frac{1+3e^{-\lambda z}}{(1+e^{-\lambda z})^2}$$

$$n_1 = \lambda^2 \mu_1 / 4\pi K .$$

(2.25)

3. Introduction of Field Fluctuations

Since we are restricting our attention to thermal equilibrium, it behooves us to make use of the well developed formalism of equilibrium statistical mechanics. According to this, momenta contribute only trivially to thermodynamic quantities, and the probability distribution in coordinate space for N particles at reciprocal temperature β is given by

$$\rho(\underline{r}_1, \ldots, \underline{r}_N) = \exp -\beta(\Phi(\underline{r}_1, \ldots, \underline{r}_N) - F)$$

where $\quad e^{-\beta F} = \int \cdots \int e^{-\beta \Phi(\underline{r}_1, \ldots, \underline{r}_N)} d\underline{r}_1 \cdots d\underline{r}_N .$

(3.1)

Φ is the full coordinate space potential and F is then the Helmholtz free energy. If the number of particles, N, is not fixed but is rather determined by a reservoir of chemical potential μ, then instead

$$\rho_N(\underline{r}_1, \ldots, \underline{r}_N) = \frac{1}{N!} \exp \beta(\Phi(\underline{r}_1, \ldots, \underline{r}_N) - N\mu - \Omega)$$

where $\quad e^{-\beta\Omega} = \sum_N \frac{1}{N!} e^{\beta N\mu} \int \cdots \int e^{-\beta\Phi(\underline{r}_1, \ldots, \underline{r}_N)} d\underline{r}_1 \cdots d\underline{r}_N .$

(3.2)

Ω is the grand potential, with the value $-PV$ for a uniform system of volume V. Finally, if the degrees of freedom include a field $\{v(\underline{r})\}$, (3.2) must be generalized to

$$\rho_N(\underline{r}_1, \ldots, \underline{r}_N; v(\underline{r})) = \frac{1}{N!} \exp -\beta(\Phi(\underline{r}_1, \ldots, \underline{r}_N; v(\underline{r})) - N\mu - \Omega)$$

where $\quad e^{-\beta\Omega} = \sum_N \frac{1}{N!} e^{\beta N\mu} \int\!\!\int \int \cdots \int e^{-\beta\Phi(\underline{r}_1, \ldots, \underline{r}_N; v(\underline{r}))} d\underline{r}_1 \cdots d\underline{r}_N Dv .$

(3.3)

Here $\int\!\!\int Dv$ denotes a functional integral over $\{v(\underline{r})\}$; the only property that we will (implicitly) use is that it is translation-invariant, but in fact it will always be associated in our discussion with an integrand that permits immediate interpretation as a Wiener integral.

We will concentrate on the grand potential Ω, since it serves as generating function for all expectations by virtue of the obvious relation

$$\delta\Omega = \langle\delta\Phi(\underline{r}_1,\ldots,\underline{r}_N;v(\underline{r}))\rangle \quad . \tag{3.4}$$

If the argument leading to (2.4) is to be believed, our particle system with core interaction ϕ_0, attractive tail $-\phi_1$ (with the convention $\phi_0(0) = \phi_1(0) = 0$), and external potential field u, can be represented by [5,6]

$$e^{-\beta\Omega[u]} = \sum_N \frac{1}{N!} \iint \int\cdots\int e^{-\frac{1}{2}\beta\Sigma\phi_0(\underline{r}_i-\underline{r}_j)} e^{-\beta\Sigma(u(\underline{r}_i)+v(\underline{r}_i)-\mu)} \\ e^{-\frac{1}{2}\beta\int v(\underline{r})\phi_1^{-1}v(\underline{r})d\underline{r}} d\underline{r}_1,\ldots,d\underline{r}_N Dv \tag{3.5}$$

(which can be defined providing that ϕ_1 is positive definite). To verify this, we perform the translation $v(\underline{r}) \to v(\underline{r}) - \phi_1\rho(\underline{r})$, $\rho(\underline{r})$ as in (2.6), converting (3.5) to

$$e^{-\beta\Omega[u]} = I_1 \sum_N \frac{1}{N!} \int\cdots\int e^{-\frac{1}{2}\beta\Sigma(\phi_0(\underline{r}_i-\underline{r}_j)-\phi_1(\underline{r}_i-\underline{r}_j))} \\ e^{-\beta\Sigma(u(\underline{r}_i)-\mu)} d\underline{r}_1\ldots d\underline{r}_N \tag{3.6}$$

where $I_1 = \iint e^{-\frac{1}{2}\beta\int v(\underline{r})\phi_1^{-1}v(\underline{r})d\underline{r}} Dv$.

Thus, except for an irrelevant constant, (3.6) is precisely the required expression for the system of interacting particles.

Our strategy however will be to carry out the integration over $\underline{r}_1,\ldots,\underline{r}_N$ and summation over N instead. Doing so, (3.5) becomes

$$e^{-\beta\Omega[u]} = \iint e^{-\beta\Omega_0[u+v]} e^{-\frac{1}{2}\beta\int v(\underline{r})\phi_1^{-1}v(\underline{r})d\underline{r}} Dv \; , \tag{3.7}$$

which will serve as our basic expression. We have in fact set up a probability density in v-space

$$\mathscr{P}_v[u] = e^{-\beta(\mathcal{H}_v[u]-\Omega[u])}$$

where

$$\mathcal{H}_v[u] = \Omega_0(u+v) + \frac{1}{2}\int v(\underset{\sim}{r})\phi_1^{-1}v(\underset{\sim}{r})d\underset{\sim}{r}$$

and $\quad e^{-\beta\Omega[u]} = \int\!\!\int e^{-\beta\mathcal{H}_v[u]} Dv$. (3.8)

From (3.4), it readily follows by choosing $\delta\Phi = Q$ that any expectation $<Q[u]>$ in the presence of the external field u can be written as

$$<Q[u]> = \int\!\!\int <Q[u+v]>_0 \mathcal{P}_v[u] Dv \, , \qquad (3.9)$$

and in particular

$$n(\underset{\sim}{r}) = \int\!\!\int n_0(\underset{\sim}{r}|u+v)\mathcal{P}_v[u] Dv \, . \qquad (3.10)$$

However, operations on the field can be used for the same purpose. For instance, applying $\delta/\delta v(\underset{\sim}{r})$ to the integrand of (3.5) causes the integral to vanish, thereby providing the relation

$$n(\underset{\sim}{r}) = -\phi_1^{-1}\bar{v}(\underset{\sim}{r}) \qquad (3.11)$$

(directly derivable from (3.8) as well), precisely as in (2.16).

To make use of (3.8), we have to evaluate $\Omega_0[u+v]$ and carry out the requisite integrations over v. Each task will be undertaken by essentially a transcription of the approximations of Section 2. To start with, we have observed that for a uniform system, $\Omega = -PV = -\int Pd\underset{\sim}{r}$. This suggests that if the reference system is not varying rapidly in space, so that the local pressure would be expected to depend upon the local excess chemical potential,

$$\Omega_0[\mu] = -\int P_0(\mu - u(\underset{\sim}{r}))d\underset{\sim}{r} \, . \qquad (3.12)$$

Indeed, since (from (3.4)) $n(\underset{\sim}{r}) = \delta\Omega/\delta u(\underset{\sim}{r})$, this implies

$$n_0(\underset{\sim}{r}) = \partial P_0(\mu - u(\underset{\sim}{r}))/\partial\mu \, , \qquad (3.13)$$

just a statement of the local thermodynamics approximation. We thus have

$$\mathcal{H}_v[\mu] = \int [\tfrac{1}{2}v(\underset{\sim}{r})\phi_1^{-1}v(\underset{\sim}{r}) - P_0(\mu - u(\underset{\sim}{r}) - v(\underset{\sim}{r}))]d\underset{\sim}{r} \, , \qquad (3.14)$$

an expression that we will retain.

We proceed next to approximations on $\mathcal{P}_v[u]$ that allow the integration Dv to be performed. Simplest is the no-fluctuation assumption: the distribution of v is concentrated at the minimum of

$\mathcal{H}_v[u]$. Thus $\bar{v}$ minimizes $\mathcal{H}_v[u]$, and from $\delta \mathcal{H}_v[u]/\delta v(\underset{\sim}{r}) = 0$, we have

$$\phi_1^{-1}\bar{v}(\underset{\sim}{r}) = -n_0(\mu - u(\underset{\sim}{r}) - \bar{v}(\underset{\sim}{r})) , \qquad (3.15)$$

subsequent to which (3.10) or (3.11) tells us that

$$n(\underset{\sim}{r}) = n_0(\mu - u(\underset{\sim}{r}) - \bar{v}(\underset{\sim}{r})) . \qquad (3.16)$$

Eq. (3.15), when $u(\underset{\sim}{r}) = 0$, is just the mean field approximation (2.16), the consequences of which we have analyzed above.

We can now sequentially improve the mean field approximation by expanding $\mathcal{H}_v[u]$ about the minimizing field $\bar{v}$. The next non-vanishing order is the second. From (3.14), we have then

$$\mathcal{H}_v[u] = \int [\tfrac{1}{2}\bar{v}(\underset{\sim}{r})\phi_1^{-1}\bar{v}(\underset{\sim}{r}) - P_0(\mu - u(\underset{\sim}{r}) - \bar{v}(\underset{\sim}{r}))]d\underset{\sim}{r}$$
$$+ \tfrac{1}{2}\int \delta v(\underset{\sim}{r})[\phi_1^{-1} - n_0'(\mu - u(\underset{\sim}{r}) - v(\underset{\sim}{r}))]\delta v(\underset{\sim}{r})d\underset{\sim}{r} , \qquad (3.17)$$

where $\delta v(\underset{\sim}{r}) = v(\underset{\sim}{r}) - \bar{v}(\underset{\sim}{r})$, converting $\mathcal{P}_v[u]$ to a Gaussian. Now from (3.10) and (3.13),

$$n(\underset{\sim}{r}) = \iint n_0(\mu - u(\underset{\sim}{r}) - \bar{v}(\underset{\sim}{r}) - \delta v(\underset{\sim}{r}))\mathcal{P}_v[u]Dv , \qquad (3.18)$$

for which only the distribution of the single quantity $\delta v(\underset{\sim}{r})$ is needed. But this is Gaussian, $\langle\delta v(\underset{\sim}{r})\rangle = 0$, and so we need only

$$\zeta(\underset{\sim}{r})^2 \equiv \langle\delta v(\underset{\sim}{r})^2\rangle = \tfrac{1}{\beta}[\phi_1^{-1} - n_0'(\mu - u(\underset{\sim}{r}) - \bar{v}(\underset{\sim}{r}))]^{-1}(\underset{\sim}{r},\underset{\sim}{r}) , \qquad (3.19)$$

as a consequence of which

$$\mu(\underset{\sim}{r}) = \int n_0(\mu - u(\underset{\sim}{r}) - \bar{v}(\underset{\sim}{r}) - \Delta)e^{-\Delta^2/2\zeta(\underset{\sim}{r})^2}d\Delta/\sqrt{2\pi\zeta(\underset{\sim}{r})^2} . \qquad (3.20)$$

In other words, the full effect of fluctuations in v at this level is that the profile that would have resulted in the absence of Δ is now broadened via a chemical potential broadening with an amplitude of $\zeta(\underset{\sim}{r})$. It is then the structure of $\zeta(\underset{\sim}{r})$ that we must examine in detail.

4. Capillary Wave Picture

With the phase interface problem in mind, let us set $u(\underset{\sim}{r}) = 0$. This however raises a problem. Taking the gradient of (3.15), with

$u(\underset{\sim}{r}) = 0$, we have

$$(\phi_1^{-1} - n_0'(n-\bar{v}(\underset{\sim}{r})))\nabla\bar{v}(\underset{\sim}{r}) = 0 \quad , \tag{4.1}$$

informing us that the inverse operator of (3.19) does not exist. Since the particle number is no longer fixed, a self-supported phase interface can be translated arbitrarily in the system with no change of controlling parameters. The null eigenvector $\nabla\bar{v}(\underset{\sim}{r})$ represents precisely such an infinitesimal translation. It would disappear under a weak applied field $u(\underset{\sim}{r})$ or under fixing of particle number, and so we shall simply imagine that (3.19) represents a principal part or generalized inverse, with the null eigenspace projected out.

Let us also anticipate a problem that will arise from the local thermodynamics approximation which, in effect, says that the correlation range of the core fluid is zero. This means that there is no built-in limitation on the rapidity with which the density can change and thus allows excitation of arbitrarily large wave vector to propagate unchecked. A semblance of reality can be recovered by instituting a cutoff K_M on wave vector space, and this is what we shall do whenever necessary.

With these caveats out of the way, what is the nature of the dispersion $\zeta(\underset{\sim}{r})$? We once more have in mind the situation of a plane symmetric interface, with tail potential modeled by (2.18). Hence it is the operator

$$\mathcal{L} = \frac{1}{4\pi K}\left(\lambda^2 - \frac{\partial^2}{\partial z^2}\right) - n_0'(\mu-\bar{v}(z)) - \frac{1}{4\pi K}\nabla_T^2$$

$$\nabla_T^2 = \frac{\partial^2}{\partial x^2} + \frac{\partial^2}{\partial y^2} \tag{4.2}$$

that we must analyze. In the special case of (2.23), the analysis can be carried out in complete detail [7]. But this is not necessary. Let us observe instead that if the system is infinite in the z-direction, but a periodic box $L \times L$ in cross-section, then $\mathcal{L}$ has the spectrum

$$\ell_{\alpha\underset{\sim}{k}} = \ell_\alpha + \frac{k^2}{4\pi K} \quad , \qquad \underset{\sim}{k} = \frac{2\pi}{L}(N_1, N_2)$$

where

$$-\frac{1}{4\pi K}\frac{\partial^2 \psi_\alpha}{\partial z^2} + \left[\frac{\lambda^2}{4\pi K} - n_0'(\mu-\bar{v}(z))\right]\psi_\alpha = \ell_\alpha \psi_\alpha \quad . \tag{4.3}$$

The lowest eigenvalue in z-space is $\ell_0 = 0$, corresponding to the nodeless state $\psi_0 = \bar{v}'(z)$, and $-n_0'(\mu-v(z))$ is asymptotically 0, with a negative well in the vicinity of the origin, so that ℓ_1 is at a finite separation above ℓ_0. Thus, there is a sequence of states

$$\psi_{ok}(\underset{\sim}{r}) = \frac{C}{L^2} \bar{v}'(z) e^{i\underset{\sim}{k}\cdot\underset{\sim}{x}}$$

$$C = \left[\int (\bar{v}'(z))^2 dz\right]^{-\frac{1}{2}} \tag{4.4}$$

$$\ell_{ok} = k^2/4\pi K ,$$

whose eigenvalues have 0 as limit point, which can make a divergent contribution to (3.19) (for $L \to \infty$):

$$\begin{aligned}
\zeta(\underset{\sim}{r})^2 &= \frac{1}{\beta} \sum_{(\alpha,k)\neq(0,0)} |\psi_{\alpha k}(\underset{\sim}{r})|^2/\ell_{\alpha k} \\
&= \frac{4\pi K}{\beta} \frac{C^2}{L^{D-1}} \bar{v}'(z)^2 \sum_{|k|>0}^{K_M} \frac{1}{k^2} + \ldots ,
\end{aligned} \tag{4.5}$$

where we have generalized from 3 to D-dimensional space.

The behavior of (4.5) under increase of the cross-section diameter L is a sensitive function of the dimensionality D of the space in which the system lives. Since $\underset{\sim}{k}$ is one dimension lower, we see that (note that K changes its dimension as D changes).

$$\zeta(\underset{\sim}{r})^2 = \begin{cases} \frac{4K}{\pi\beta} C^2 \bar{v}'(z)^2 L + \ldots & D = 2 \\ \frac{K}{\pi\beta} C^2 \bar{v}'(z)^2 \ln(K_M L) + \ldots & D = 3 \\ \frac{2K}{\pi\beta} C^2 \bar{v}'(z)^2 K_M + \ldots & D = 4 \end{cases} \tag{4.6}$$

Hence the interface of a 2-dimensional fluid rapidly broadens — as $L^{\frac{1}{2}}$ — with increasing interfacial area; for a 3-dimensional fluid, the broadening is much slower, and it is first in 4 dimensions that a density profile independent of L can be achieved in the usual thermodynamic limit of infinite volume.

This conclusion, that fluctuations — field fluctuations in our picture — prohibit the establishment of a volume-independent density profile in two or three dimensions, is not new. It has usually been obtained in the context of capillary waves [8], bulk motions of the interfacial region. Transcription to this picture is not difficult.

We return to (3.18) and note that if

$$n_0(\mu - \bar{v}(z)) = \bar{n}(z) \tag{4.7}$$

is the mean field approximation profile, then to first order in Δ _in the argument_,

$$\bar{n}(z - \Delta) = n_0(\mu - \bar{v}(z) + \bar{v}'(z)\Delta) . \tag{4.8}$$

Thus, (3.18) can be rewritten

$$n(\underset{\sim}{r}) = \iint \bar{n}(z - \Delta) \mathcal{P}_{\bar{v}-\bar{v}'(z)\Delta}[u] Dv . \tag{4.9}$$

If the distribution is now taken as Gaussian in Δ, (2.20) changes to

$$n(\underset{\sim}{r}) = \int \bar{n}(z-\Delta) e^{-\Delta^2/2\xi(z)^2} d\Delta / \sqrt{2\pi \xi(z)^2}$$

where (4.10)

$$\xi(z)^2 = \left\langle \left(\frac{\delta v(z)}{\bar{v}'(z)}\right)^2 \right\rangle = \frac{1}{\beta} [\phi_1^{-1} - n_0'(\mu - \bar{v}(z))]^{-1} (\underset{\sim}{r},\underset{\sim}{r}) / \bar{v}'(z)^2 .$$

But then according to (4.6), $\bar{v}'(z)^2$ cancels completely in (4.10), e.g. in three dimensions

$$\xi(\underset{\sim}{r})^2 = \frac{KC^2}{\pi \beta} \ln(K_M L) + - \tag{4.11}$$

is independent of z. It thus represents a bodily motion of the whole interface at each location in the transverse plane — in other words, a capillary motion. However, we are left with the feeling that the notion of a force field tethered to a disordered gas is more basic, more controllable, and more amenable to meaningful approximation.

---

[*]Supported in part by the Department of Energy, Contract DE-AC02-76ERO3077 and National Science Foundation, Grant CHE-80011285.

REFERENCES

1. J.D. Van der Waals, Z.Phy. Chem. 13, 657 (1894).

2. J.K. Percus, Courant Institute Course Notes, "Many-Body Theory," 1965 (unpublished).

3. J.K. Percus, Trans. N.Y. Acad. Sci. 26, 1062 (1964).

4. N.G. Van Kampen, Phys. Rev. 135, A 362 (1964).

5. M. Kac, in "Applied Probability," ed. L.A. MacColl (McGraw-Hill, New York 1957).

6. A.J.F. Siegert, in "Statistical Physics," 1962 Brandeis Lecture (Benjamin, New York, 1963).

7. J.K. Percus, in "Studies in Statistical Mechanics," ed. J.L. Lebowitz and E.W. Montroll (North-Holland, Amsterdam 1981).

8. Buff, Lovett, and Stillinger, Phys. Rev. Lett. 15, 621 (1965).

FINGERING IN POROUS MEDIA

P. G. Saffman
Department of Applied Mathematics
California Institute of Technology
Pasadena, CA 91125

Microscopic Fingering

Fingering is the phenomena which occurs when an attempt is made to displace a fluid saturating a porous medium by injecting another immiscible fluid of smaller viscosity. In this case, it is well known that the injected fluid tends to advance by sending out fingers and a fraction of the original fluid is left behind in the pores and interstices of the medium. A problem of major interest is to understand how the amount left behind and its spatial distribution depends upon the properties of the fluids and the medium.

One difficulty is that the fingering occurs on two scales, the microscopic and the macroscopic. The microscopic scale describes the motion of the interface in the pores and crevices. Unless the speeds are high, the viscosity small or the porosity large, the viscous forces dominate and the problem is essentially that of low Reynolds number (Stokes) flow of two immiscible fluids in a region of irregular geometry. The real problem is impossibly hard, but the fluid dynamical processes can be exemplified by motion in a capillary tube, where it is imagined the more viscous fluid wets the walls and that a finger of the less viscous fluid advances along the axis of the tube displacing the original more viscous fluid but leaving a film of it adhering to the wall. This flow is not too difficult to set up experimentally and experiments were carried out by Fairbrother & Stubbs (1935) and more extensively by Taylor (1961) and Cox (1962). On dimensional grounds, it is expected that the fraction of viscous fluid left behind, m say, is a function only of $\mu u/T$ and μ/μ_1, where u is the velocity of the tip of the finger, μ is the viscosity of the displaced fluid, μ_1 is the viscosity of the displacing fluid, and T is the interfacial tension. Since the inertia forces are assumed negligible and the flow is governed by a balance between viscous and surface tension forces, the size of the capillary does not enter and

$$m = F\left(\frac{\mu u}{T}, \frac{\mu_1}{\mu}\right). \tag{1}$$

The experiments were carried out with $\mu_1/\mu \ll 1$ and the form of F for this case measured. In the limits of small and large values of $\mu u/T$,

$$m \doteq \left(\frac{\mu u}{T}\right)^{1/2} \quad \text{for} \quad \frac{\mu u}{T} < 0.1, \qquad (2)$$

and

$$m \doteq 0.6 \quad \text{for} \quad \frac{\mu u}{T} > 2. \qquad (3)$$

The calculation of F from first principles using the Stokes (creeping flow) equations is a problem of considerable difficulty. Bretherton (1961) attempted to tackle the $\mu u/T \ll 1$ limit, in which the thickness of the film of viscous fluid is small, by matching a lubrication flow solution for the film shape with the approximately constant curvature shape over the front. However, the analysis gives

$$m \propto \left(\frac{\mu u}{T}\right)^{2/3}, \qquad (4)$$

and the conflict between the experiments and the asymptotic calculation has still not apparently been resolved.

An attempt is currently under way at the California Institute of Technology to solve the problem for finite $\mu u/T$ by numerical solution of the governing equations. There is a serious question about the uniqueness of the mathematical problem, since recent work has uncovered numerous examples of non-uniqueness of flows with free surfaces (e.g. Chen & Saffman 1980), and it is hoped that the work may give some information about this matter. In the event that non-unique solutions are obtained, we shall be faced with the serious problem of deciding what determines the actual value of m. This investigation is, however, still in a preliminary stage and no results are yet available.

If the porous medium were just a bundle of straight parallel capillary tubes, then the fingering would be entirely microscopic and the amount left behind described by (1), but of course this is a gross oversimplification. An attempt was made by Saffman (1963, 1969) to incorporate (1) into a more realistic (but still oversimplified) model of a porous medium as a random collection of capillary tubes, inclined at random directions with equal probability. Then the amount left behind by microscopic fingering in a capillary inclined at an angle θ to the direction of the mean flow is

$$F\left(\frac{3\mu U \cos \theta}{T}, \frac{\mu_1}{\mu}\right) \qquad (5)$$

where U is the 'average velocity' of the interfaces or the velocity of the 'average interface'. However, (5) is predicated on the assumption that the finger goes all the way from one end of the capillary to the other. In fact, fluid will be trapped in a capillary because the less viscous fluid gets to the far end through some other capillary before the finger gets there. In this way, a fraction m_t will be trapped in the capillaries by purely kinematic effects, independent of the average speed U, and if $\mu_1 \ll \mu$ this fluid will in fact be trapped permanently. Making the ad hoc but reasonable guess that the proportion trapped in a capillary is on average proportional to $(1 - \cos\theta)$, one obtains the expression for the total fraction left behind

$$m = m_t + \frac{T}{3\mu U} \int_0^{\frac{3\mu U}{T}} F\left(\xi, \frac{\mu_1}{\mu}\right) \left(1 - 2m_t + \frac{2m_t T \xi}{3\mu U}\right) d\xi, \qquad (6)$$

where m_t is a function only of the geometry, and $F(\xi, \mu_1/\mu)$ can be determined from experiments in straight capillaries and also, it is to be hoped, eventually from numerical solutions. For small values of $\mu U/T$,

$$m \doteq m_t + \frac{2}{\sqrt{3}}\left(1 - \frac{4m_t}{5}\right)\left(\frac{\mu U}{T}\right)^{\frac{1}{2}} \qquad (7)$$

whereas for large values of $\mu U/T$,

$$m \doteq 0 \cdot 6 + 0 \cdot 4 \, m_t . \qquad (8)$$

The extent to which this result models the real phenomena is not known, as suitable experiments with which to compare are not known to the author. One reason why experiment is not simple and it may be difficult to interpret the results unambiguously is that the distribution of fluid may also be affected by macroscopic fingering on a scale large compared to the pore size.

Macroscopic Fingering

The above description of microscopic fingering gives a picture of the flow divided into two regions, one occupied by a single phase of viscous fluid and the other a mixture of the two phases. It is reasonable to suppose that under certain conditions the transition region, in which variations of microscopic finger are dominant, is thin compared with some overall macroscopic dimension of the flow and the flow can be modelled as composed of two macroscopic domains, separated by an interface, in each of which an appropriate form of Darcy's law is satisfied. The fraction m of original fluid in the region behind the interface is then regarded, in the spirit of Darcy's law, as a continuous function of position and time, with a discontinuity at the interface

$$m = \bar{F}\left(\frac{\mu U_n}{T}, \frac{\mu_1}{\mu}\right) \tag{9}$$

where U_n is the normal velocity of the interface and $\bar{F}$ depends also on the porosity, permeability, etc., and also on the wetting properties of the fluids.

An important question is under what conditions is it reasonable to assume a macroscopic **interface**. The author can at present offer no satisfactory answer to this question, other then the speculation that $s = \mu U k^{\frac{1}{2}}/LT$, where k is the permeability and L the macroscopic length scale, measures the fuzziness of the interface produced by microscopic fingering and $s \ll 1$ is sufficient. A suitable name for s is the sharpness parameter. Note that it is of course possible to have $\mu U/T = O(1)$ and $s \ll 1$, if the permeability is sufficiently small.

Macroscopic fingering can be thought of as the finite amplitude stage of the so-called Saffman-Taylor instability of the plane interface in a porous medium separating two fluids with the less viscous driving the more viscous fluid. As shown by Saffman & Taylor (1958), the case in which $m = $ constant is mathematically equivalent to the case $m = 0$, and for theoretical calculations it is therefore sufficient to suppose that one fluid completely expels the other if the dependence of m on U is negligible. Exact solutions for two-dimensional flow, based on the further assumption that the pressure drop across the interface is constant, can be obtained which describe both the growth of infinitesimal disturbances to finite size (Saffman 1959) and the shapes of steady fingers (Saffman & Taylor 1958). Although, these two-dimensional solutions are unlikely to be directly relevant to porous media, where it is expected that the fingers will be macroscopically three-dimensional, they can be tested experimentally in the device known as the Hele-Shaw cell, which is the space between two parallel, closely spaced flat plates. The average velocity in the gap is

$$u = -\frac{b^2}{12\mu}\frac{\partial p}{\partial x}, \qquad v = -\frac{b^2}{12\mu}\frac{\partial p}{\partial x} \tag{10}$$

where b is the distance between the plates, and the equations are those of a two-dimensional porous medium with permeability $b^2/12$ obeying Darcy's law.

Experiments in Hele-Shaw cells show good qualitative agreement with the predictions of the theory (see plates 2 and 3 of Saffman & Taylor 1958) and excellent quantitative agreement between observed and measured shapes is found for the faster bubbles (for which the assumption of constant interfacial pressure drop is most reasonable) provided the finger width is taken from the data. For some recent observations on radial fingering associated with injection or extraction from a point source in a Hele-Shaw cell, see Paterson (1981). There are, however, serious problems which arise from a further development of the theory and the attempt to predict the width of the finger, which indicate that although it may be reasonable under certain (but not yet well established) conditions to suppose that the microscopic fingering is restricted to a mathematically sharp interface, yet the boundary conditions on the macroscopic (Darcy) flow depend on the structure of the interface and plausible assumptions about the jump conditions lead to results in contradiction with experiment. In view of the several recent and proposed numerical calculations of unsteady macroscopic interfaces in porous media (e.g. Meng & Thompson 1978, Glimm et al 1980) it cannot be stressed too strongly that the question of the boundary conditions needs to be addressed and settled before the calculations can be regarded as more than an academic exercise.

The Hele-Shaw cell is an excellent phenomenon to study in this connection, as it contains a great deal of the physics without the complication of the irregular or random geometry. Suppose we consider a finger moving in a channel in a Hele-Shaw cell. The parallel plates are the surfaces $z = \pm \frac{1}{2}b$, and the edges of the channel are $y = \pm a$. For simplicity, suppose that $\mu_1 = 0$ (e.g. a finger of air displaces a heavy oil). In the region external to the finger, the average velocity $\underset{\sim}{u}_E(x,y,t)$ satisfies

$$\underset{\sim}{u}_E(x,y,t) = -\frac{b^2}{12\mu} \nabla p_E, \quad \text{div } \underset{\sim}{u}_E = 0 . \tag{14}$$

In the region occupied by the finger, there is a film of viscous fluid of total thickness mb, whose average velocity satisfies the equation

$$\underset{\sim}{u}_I (x,y,t) = -\frac{m^2 b^2}{12\mu} \nabla p_I , \tag{15}$$

together with the equation of continuity

$$\frac{\partial m}{\partial t} + \text{div}(m \underset{\sim}{u}_I) = 0 . \tag{16}$$

In general, $m = m(x,y,t)$. The remaining equation is

$$\tfrac{1}{2} b T \nabla^2 m + p_I = 0 \tag{17}$$

which expresses continuity of force across the surfaces of the finger parallel to the plates.

The boundary conditions at the edge of the finger are

$$(\underset{\sim}{U} - \underset{\sim}{u}_E) \cdot \underset{\sim}{n} = m (\underset{\sim}{U} - \underset{\sim}{u}_I) \cdot \underset{\sim}{n} \tag{18}$$

where $\underset{\sim}{U}$ is the velocity of the finger and $\underset{\sim}{n}$ is the normal direction,

$$m = F (\mu \underset{\sim}{U} \cdot \underset{\sim}{n}/T) \tag{19}$$

and

$$p_E = p_I + \frac{T}{b} f\left(\frac{\mu \underset{\sim}{U} \cdot \underset{\sim}{n}}{T}, \frac{b}{R}\right) \tag{20}$$

where R is the radius of curvature of the edge in the x-y plane. The functions F and f are at present unknown and are to be determined by the 'microscopic' conditions at the edge. In the porous medium case, they would result from studies of the microscopic fingering process. In the Hele-Shaw flow, they can in principle be found from solutions of the Stokes equations in the $n - z$ plane. (Actually, Glimm et al 1980 assume for their calculations that $F = 0$, i.e. that the saturation m is continuous across the interface. This is equivalent to supposing that there is no difference between microscopic and macroscopic fingering, other than a change of scale, and that the variations on the microscopic scale can be described adequately by macroscopic equations. For porous media in which there is a large and continuous gradation of pore sizes, this approach may be a good approximation, but it does not seem to be a satisfactory representation of interfacial motion in a Hele-Shaw cell and is probably not good if the distribution of pore sizes in the medium is sharp).

Steady solutions of these equations and their linear stability are currently under study with plausible forms assumed for F and f. They have been solved (McLean & Saffman 1981) for the case

$$F = \text{const.}, \quad f = f_o + c \frac{b}{R}. \qquad (21)$$

The analysis of Saffman & Taylor (1958) is the special case $c = 0$. The steady solutions are then arbitrary and the asymptotic width $2\lambda a$ of the finger could not be determined. However, taking $\lambda = \frac{1}{2}$ gave shapes which were in excellent agreement with the observed shapes for faster fingers. Slower fingers were observed to have values of $\lambda > \frac{1}{2}$, which depend on $\mu U/T$, and shapes which do not conform to those calculated for these values of λ. The inclusion of the term involving c removes the arbitrariness and gives λ as a unique function of $\mu UT\, b^2/a^2$, which agrees reasonably with the Saffman-Taylor experiments, and those of Pitts (1980) if $c \approx 3-4$. The observed shapes are in excellent agreement with calculation when the asymptotic widths are fixed equal.

However, the apparent success of the theory employing (21) is probably illusory. Apart from the fact that the value of c is somewhat large, a value of 1 being much more plausible, a study of the stability of the fingers reveals that they are exponentially unstable to infinitesimal disturbances. The observations show that they are extremely stable, without a trace of unsteadiness, and the conclution must be that the boundary condition (21) are wrong, at least for unsteady motion. A further test would be to carry out experiments for different values of a/b.

We conclude that microscopic and macroscopic fingering are strongly coupled, and although it is too much to hope for a complete theoretical solution in the foreseeable future, research should at least uncover the empirical dependence.

Acknowledgement

This work was supported by the Department of Energy (Office of Basic Energy Sciences). Thanks are also due to Control Data Corporation for the granting of time on the CYBER 203 at the C.D.C. Service Center, Arden Hills, Minnesota.

References

Bretherton, F. P. 1961. The motion of long bubbles in tubes. J. Fluid Mech. <u>10</u>, 166-188.

Chen, B. & Saffman, P. G. 1980. Steady gravity-capillary waves on deep water 2. Numerical results for finite amplitude. Studies in App. Math. <u>62</u>, 95-112.

Cox, B. G. 1962. On driving a viscous fluid out of a tube. J. Fluid Mech. <u>14</u>, 81-96.

Fairbrother, F. & Stubbs, A. E. 1935. The 'bubble tube' method of measurement. J. Chem. Soc. <u>1</u>, 527-530.

Glimm, J., Marchesin, D. & McBryan, O. 1980. Statistical fluid dynamics: unstable fingers. Commun. Math. Phys. <u>74</u>, 1-13.

Meng, J. C. S. & Thomson, J. A. L. 1978. Numerical studies of some nonlinear hydrodynamic problems by discrete vortex element methods. J. Fluid Mech. <u>84</u>, 433-453.

Paterson, L. 1981. Radial fingering in a Hele-Shaw cell. J. Fluid Mech. (to appear).

Pitts, E. 1981. Penetration of fluid into a Hele-Shaw cell. J. Fluid Mech. <u>97</u>, 53-64.

Saffman, P. G. 1959. Exact solutions for the growth of fingers from a flat interface between two fluids in a porous medium or Hele-Shaw cell. Qu. J. Mech. App. Math. <u>12</u>, 146-150.

Saffman, P. G. 1963. The displacement of a viscous fluid from a porous medium. Proceedings 1963 Heat Transfer and Fluid Mechanics Institute p176-182. Stanford U. Press.

Saffman, P. G. 1969. A mathematical treatment of dispersion in flow through a branching tree. Proc. CIBA Symposium on Circulatory and Respiratory Mass Transport. (Ed. Wolstenholme & Knight). Churchill.

Saffman, P. G. & Taylor, G. I. 1958. The penetration of a fluid into a porous medium or Hele-Shaw cell containing a more viscous liquid. Proc. Roy. Soc. A<u>245</u>, 312-329.

Taylor, G. I. 1961. Deposition of a viscous fluid on the wall of a tube. J. Fluid Mech. <u>10</u>, 161-165.

ON THE EFFECTIVE THERMAL CONDUCTIVITY AND

PERMEABILITY OF REGULAR ARRAYS OF SPHERES

A.S. Sangani and A. Acrivos
Department of Chemical Engineering
Stanford University
Stanford, CA 94305

Abstract

We consider two-phase materials in which the discrete phase consists of equal-sized spheres fixed in a periodic array and summarize the results for the effective thermal conductivity and the permeability of such materials. The results are given for the whole range of volume fractions of spheres for three cubic arrays: simple, body-centered and face-centered.

1. Introduction

The problem of determining the effective parameters, e.g. conductivity, permeability, etc. of macroscopically homogeneous two phase materials has received a great deal of attention owing to its importance in many branches of mathematical physics. Here we shall be concerned with one very special aspect of this field, viz. when the discrete phase consists of equal-sized spheres fixed in a regular array. This renders the problem deterministic and therefore amenable to a unique solution. In spite of their specialized nature, the solutions to be presented provide valuable information which, at least qualitatively speaking, is applicable to more general systems. Also the method of solution which we have employed is of interest in its own right and might be found useful in many other applications.

We shall consider separately the two problems of thermal conduction and of flow through a periodic array.

2. The Effective Thermal Conductivity

Consider a composite material consisting of spherical particles of radius a and conductivity α embedded in a homogeneous matrix of unit conductivity. Rayleigh was the first to analyze the case when the spheres are arranged in a simple cubic array. He developed a solution by replacing the spheres by dipoles and higher order multipoles and obtained for the effective thermal conductivity k^*:

$$k^* = 1 - 3c \left[\frac{2+\alpha}{1-\alpha} + c - \frac{(1-\alpha)d}{(4+3\alpha)} c^{10/3} + O(c^{14/3}) \right]^{-1} \qquad (1)$$

where c is the volume fraction of the solids. The value $d = 4.95$ given by Rayleigh was later corrected by Runge [9] to 1.57. Meredith and Tobias [7] extended Rayleigh's analysis and calculated the coefficient of the $O(c^{14/3})$ term. The validity of Rayleigh's method was questioned in recent years (see for example, Levine [4]; Jeffrey [3]) because it involved the summation of a non-absolutely convergent series. O'Brien [8] and McPhedran and McKenzie [6] have, however, modified Rayleigh's method in order to overcome this difficulty, hence, this modified Rayleigh's method has now a sound theoretical basis. McPhedran and McKenzie also gave an expression for k^* to $O(c^{35/3})$. Unfortunately, their calculated coefficients are somewhat incorrect because they were obtained using an axially symmetric potential, rather than a potential possessing the four-fold symmetry of the cubic array.

Recently Zuzovski and Brenner [13] resolved this problem using the method of generalized functions, which avoids the difficulty encountered in Rayleigh's original treatment, and found that the coefficient of the $O(c^{14/3})$ term reported by Meredith and Tobias [7] was in error. Unfortunately, owing to a numerical slip, the "corrected coefficient" given by Zuzovski and Brenner is itself incorrect.

We have modified the approach taken by Zuzovski and Brenner [13] and have derived expressions for k^* to $O(c^9)$ for any α for simple, body-centered, and face-centered cubic arrays. We have also calculated k^* over the complete range of volume fractions of solids for the two special cases: $\alpha = \infty$ and $\alpha = 0$. Our numerical results for $\alpha = \infty$ are indistinguishable from McPhedran and McKenzie's [6] for simple cubic arrays, and from McKenzie, McPhedran and Derrick's [5] for body-centered and face-centered cubic arrays.

At the other extreme, i.e. for closely packed arrays, Batchelor and O'Brien [1] used lubrication theory and obtained, for perfectly conducting solids ($\alpha = \infty$), the expression

$$k^* \sim -K_1 \ln(1-\chi) - K_2 \qquad c \rightarrow c_{max}, \qquad (2)$$

where $\chi = (c/c_{max})^{1/3}$ with c_{max} being the volume fraction of the particles when they are touching each other. The constant K_1 equals $\pi/2$, $\sqrt{3}\pi/2$ and $\sqrt{2}\pi$ for, respectively, simple, body-centered, and face-centered cubic arrays for which the corresponding values for c_{max} are $\pi/6$, $\sqrt{3}\pi/8$, and $\sqrt{2}\pi/6$. On the other hand, the $O(1)$ constant K_2 in (2) cannot be determined from lubrication theory alone. By extending the series for k^* to $O(c^{15})$ for $\alpha = \infty$ we have been able to compute K_2 for the three cubic arrays which we found to equal 0.7, 2.4 and 7.2 respectively.

The method of solution, which is described in more detail elsewhere [10] is as follows:

Let the centers of the spherical particles be located at

$$\underline{r_n} = n_1 \underline{a}(1) + n_2 \underline{a}(2) + n_3 \underline{a}(3) \qquad (n_1, n_2, n_3 = 0, \pm 1, \pm 2, \cdots), \qquad (3)$$

where $\underline{a}_{(1)}$, $\underline{a}_{(2)}$, and $\underline{a}_{(3)}$ are the basic vectors determining the unit cell of the array. Further suppose that a temperature gradient in the x_1-direction of unit magnitude is applied by some external means to this composite material. We are interested in calculating along x_1 the average heat flux which gives us directly k^*.

We recall that the temperature T is a continuous harmonic function everywhere within the composite except at the surface of the particles where its normal derivative suffers a jump discontinuity if α differs from unity. This jump in the derivative is such that the heat flux is continuous everywhere. The gradient of the temperature is a periodic function and since a temperature gradient of unit magnitude is externally applied in the x_1-direction, the temperature can be expressed as

$$T = x_1 + \tilde{T} , \tag{4}$$

where $\tilde{T}$ is a spatially periodic function

$$\tilde{T}(\underline{r}) = \tilde{T}(\underline{r} + \underline{r}_n) \tag{5}$$

with $\underline{r}_n$ given by (3). In order to solve for $\tilde{T}$, we first obtain the periodic fundamental singular solution to Laplace's equation which satisfies

$$\Delta S_1(\underline{r}) = \frac{4\pi}{\tau_o} - 4\pi \sum_n \delta(\underline{r} - \underline{r}_n) \qquad \left(\Delta = \frac{\partial^2}{\partial x_1^2} + \frac{\partial^2}{\partial x_2^2} + \frac{\partial^2}{\partial x_3^2} \right) , \tag{6}$$

where τ_o is the volume of a unit cell and $\delta(\underline{r} - \underline{r}_n)$ is the Dirac's delta function defined by

$$\int_\tau \delta(\underline{r} - \underline{r}_n) d\underline{r} = \begin{cases} 1 & \text{when } \underline{r}_n \in \tau \\ 0 & \text{when } \underline{r}_n \notin \tau \end{cases} \tag{7}$$

and

$$\delta(\underline{r} - \underline{r}_n) = 0 \qquad \text{for } \underline{r} \neq \underline{r}_n . \tag{8}$$

The constant $4\pi/\tau_o$ in (6), which represents the strength of uniformly distributed sinks, must be added to the right-hand side of the equation in order to ensure that the total heat generated within a unit cell is zero.

As shown by Hasimoto [2] a solution to (6) is

$$S_1 = \frac{1}{\pi \tau_o} \sum_{\underline{k}_n \neq \underline{0}} \frac{e^{-2\pi i (\underline{k}_n \cdot \underline{r})}}{k_n^2} \tag{9}$$

where
$$\underline{k}_n = n_1 \underline{b}_1 + n_2 \underline{b}_2 + n_3 \underline{b}_3 \tag{10}$$

are the reciprocal lattice vectors given by

$$\underline{k}_n \cdot \underline{a}_{(j)} = n_{(j)} \quad (j = 1, 2, 3). \tag{11}$$

The basic vectors $\underline{b}_{(1)}$, $\underline{b}_{(2)}$, and $\underline{b}_{(3)}$ in the reciprocal lattice are related to the basic vectors in physical space by

$$\underline{b}_1 = \frac{\underline{a}_{(2)} \cdot \underline{a}_{(3)}}{\tau_o} \,, \quad \underline{b}_{(2)} = \frac{\underline{a}_{(3)} \cdot \underline{a}_{(1)}}{\tau_o} \,, \quad \underline{b}_{(3)} = \frac{\underline{a}_{(1)} \cdot \underline{a}_{(2)}}{\tau_o} \,. \tag{12}$$

The temperature field outside any sphere can now be expressed as a sum of the successive derivatives of S_1 multiplied by some unknown coefficients, i.e.

$$T = x_1 + \underline{G} \, S_1 \tag{13}$$

where $\underline{G}$ is the differential operator:

$$\underline{G} = \sum_{m+n+p=1}^{\infty} A_{mnp} \frac{\partial^{m+n+p}}{\partial x_1^m \, \partial x_2^n \, \partial x_3^p} \,. \tag{14}$$

This formal solution (13,14) is identical to that derived by Zuzovski and Brenner [13] using generalized functions. The above expression for T is singular at $\underline{r} = \underline{r}_n$ and therefore is valid only outside the particles. Within the latter, a different expression for T is needed which is regular everywhere. The requirements of continuity of temperature and flux at the surface of the particles then lead to a relation for determining these unknown coefficients A.

On using certain properties of the spherical harmonics and rearranging (14) into a more convenient form we have been able to derive a set of linear equations for the unknowns A in (14) whose coefficients we determined via an explicit expression. This set of linear equations was solved either by successive approximations (for $\chi \ll 1$) as a series in χ or by the method of "direct substitution" [10] for selected values of χ.

Table 1 gives k^*, computed via the direct substitution method, as a function of χ for perfectly conducting spheres ($\alpha = \infty$) for each of the three cubic arrays. Also, Figure 1 depicts k^* versus χ for simple cubic arrays with $\alpha = 0$ and $\alpha = \infty$. For $\alpha = \infty$, it was found that when $\chi < 0.9$ both methods converged to the same answer, but that for $0.9 < \chi < 1.0$ the series method appeared superior in that K_2 in (2) rapidly approached a constant value as the number of terms in the series increased. For simple, body-centered and face-centered cubic arrays K_2 thus calculated equaled 0.7, 2.4 and 7.2, respectively. In contrast, for $\alpha < 40$, the direct substitution method converged more quickly and the results for the case of touching spheres ($\chi = 1$) are shown in Table 2. Of particular interest are the values of k^* for $\alpha = 0$ ($\chi = 1$) which equals 0.344, 0.220, and 0.164, respectively, for the above three cubic arrays. It should be noted that these values are within 15% of the Haskin-Shrikman upper bounds for such systems.

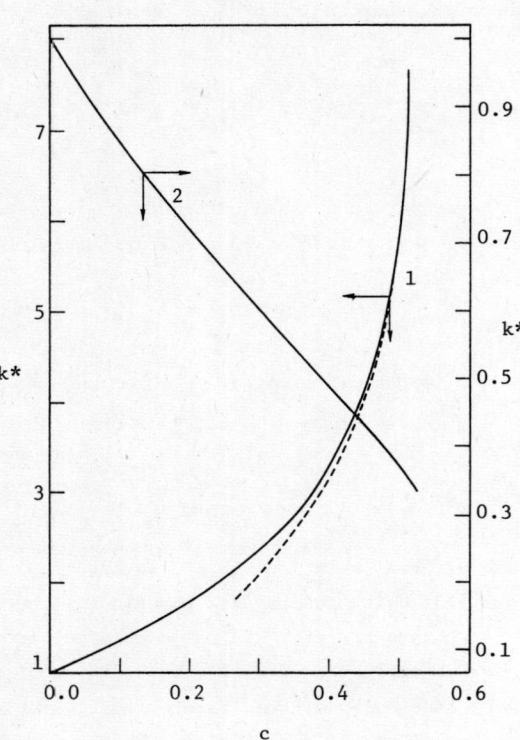

Figure 1.

The effective conductivity k^* as a function of volume fraction c for a simple cubic array of perfectly conducting (1) and perfectly insulating (2) spheres (———— computed values, ---- Eq. (2) with $K_1 = \pi/2$ and $K_2 = 0.7$) [10].

3. The Effective Permeability

Here we consider the corresponding problem of slow flow past a periodic array of spheres and wish to determine the permeability or more specifically the drag force F exerted by the fluid moving with average speed U on a representative sphere in the assembly as a function of c, the volume fraction of the spheres. As before, the centers of the spheres are located at $\underline{r}_n$ - c.f. (3).

Hasimoto [2] showed that in case of simple cubic array

$$K^{-1} = 1 - 1.7601 \, c^{1/3} + c - 1.5593 \, c^2 + 0(c^{8/3}) \tag{15}$$

TABLE 1

k* as a function of α for the three cubic arrays and for α=∞ [10]

χ	SC	BCC	FCC
0.1	1.00157	1.00204	1.00222
0.2	1.0216	1.0164	1.0179
0.3	1.0430	1.0561	1.0612
0.4	1.104	1.1365	1.1492
0.5	1.210	1.2788	1.3060
0.6	1.383	1.5168	1.5713
0.7	1.660	1.915	2.023
0.8	2.125	2.618	2.848
0.85	2.491	3.199	3.556
0.90	3.045	4.11	4.706
0.95	4.06	5.8	6.99
0.97	4.84	-	8.74
0.99	6.55	-	-

TABLE 2

k* as a function of α for various closely packed arrays (χ=1) [10]

α	SC	BCC	FCC
0	0.344	0.220	0.164
1	1.0	1.0	1.0
2	1.4585	1.6196	1.6874
5	2.422	3.026	3.338
10	3.474	4.60	5.29
20	4.81	6.45	7.68
30	-	7.6	9.1
40	6.41	8.2	10.0

where

$$K \equiv \frac{F}{6\pi\mu U a}$$

with μ being the viscosity of the fluid and a, as before, the radius of the spheres. Clearly, (15) is meaningless for c beyond approximately 0.2 when K becomes negative.

Here we modify Hasimoto's analysis and calculate K over the complete range of c for all three of the cubic arrays. Our method of solution, which is described in more detail elsewhere [11] is as follows

A periodic fundamental solution to the creeping flow equations can be obtained by solving

$$\mu \Delta v_i = \frac{\partial q}{\partial x_i} + \delta_{i1} \sum_{\underline{n}} \delta(\underline{r} - \underline{r}_n) , \quad \frac{\partial v_i}{\partial x_i} = 0 . \qquad (16)$$

As shown by Hasimoto (1959)

$$v_i = v_0 \delta_{i1} - \frac{1}{4\pi\mu} \left\{ S_1 \delta_{i1} - \frac{\partial^2 S_2}{\partial x_1 \partial x_i} \right\} \qquad (17)$$

$$\frac{\partial q}{\partial x_i} = - \frac{\delta_{i1}}{\tau_0} + \frac{1}{4\pi} \frac{\partial^2 S_1}{\partial x_1 \partial x_1} , \qquad (18)$$

where τ_0 is, as before, the volume of the unit cell, S_1 is given by (9) and

$$S_2 = \frac{-1}{4\pi^3 \tau_0} \sum_{\underline{k}_n \neq \underline{0}} \frac{e^{-2\pi i (\underline{k}_n \cdot \underline{r})}}{k_n^4} . \qquad (19)$$

S_1 satisfies (6) while

$$\Delta S_2 = S_1 . \qquad (20)$$

To obtain a completely general solution for the velocity u_i and for the pressure p, we add to the fundamental solution the derivatives of v_i and S_1 multiplied by some unknown coefficients, but omit from the sum those terms which do not satisfy the symmetry conditions of the system. We thus arrive at:

$$u_1 = U_0 - \frac{1}{4\pi\mu} \left\{ \underline{G} \left(S_1 - \frac{\partial^2 S_2}{\partial x_1^2} \right) + \underline{H} \frac{\partial^2 S_1}{\partial x_1^2} - \underline{L} \left(\frac{\partial^4}{\partial x_1^4} - 6 \frac{\partial^4}{\partial x_2^2 \partial x_3^2} + \frac{\partial^4}{\partial x_3^4} \right) S_1 \right\} \quad (21)$$

$$u_2 = \frac{1}{4\pi\mu} \left\{ \underline{G} \frac{\partial^2 S_2}{\partial x_1 \partial x_2} - \underline{H} \frac{\partial^2 S_1}{\partial x_1 \partial x_2} - \underline{L} \frac{\partial}{\partial x_1} \left(\frac{\partial^3}{\partial x_2^3} - 3 \frac{\partial^3}{\partial x_3^2 \partial x_2} \right) \right\} S_1 \quad (22)$$

$$u_3 = \frac{1}{4\pi\mu} \left\{ \underline{G} \frac{\partial^2 S_2}{\partial x_1 \partial x_3} - \underline{H} \frac{\partial^2 S_1}{\partial x_1 \partial x_3} - \underline{L} \frac{\partial}{\partial x_1} \left(\frac{\partial^3}{\partial x_3^3} - 3 \frac{\partial^3}{\partial x_2^2 \partial x_3} \right) S_1 \right\} \quad (23)$$

$$\frac{\partial p}{\partial x_i} = \frac{-F}{\tau_o} \delta_{i1} + \frac{1}{4\pi} \underline{G} \frac{\partial^2 S_1}{\partial x_1 \partial x_i} , \quad (24)$$

where $\underline{G}$, $\underline{H}$, and $\underline{L}$ are the differential operators

$$\left\{ \begin{array}{c} \underline{G} \\ \underline{H} \\ \underline{L} \end{array} \right\} = \sum_{M=0}^{\infty} \sum_{m=0}^{m \leq \frac{1}{2}M} \left\{ \begin{array}{c} A_{nm} \\ B_{nm} \\ C_{nm} \end{array} \right\} \left\{ \frac{\partial^{2n}}{\partial x_1^{2n}} \left[\left(\frac{\partial}{\partial \xi} \right)^{4m} + \left(\frac{\partial}{\partial \eta} \right)^{4m} \right] \right\} (M = n+2m) \quad (25)$$

with $\xi = x_2 + ix_3$, $\eta = x_2 - ix_3$, (26)

and where the unknown coefficients A_{nm}, B_{nm}, and C_{nm} are to be determined by applying the no-slip boundary condition at the surface of the spheres. The above expressions for the components of the velocity differ from those given by Hasimoto [2] in two important aspects. First, the terms containing the differential operator $\underline{L}$ are absent in Hasimoto's solution, and second, the differential operators defined by (26) are a special form of those given by Hasimoto and are particularly convenient when dealing with problems involving cubic arrays of spheres.

Since it has been shown by Hasimoto [2] that

$$A_{oo} = 3\pi\mu UaK \qquad U_o = U + \frac{2 B_{oo}}{\pi\tau_o}, \qquad (27)$$

the determination of A_{oo} directly yields K.

As was the case with the conductivity problem described earlier in §2, using certain properties of the spherical harmonics we were able to derive a set of linear equations for the unknown coefficients in (25). This set was then solved either by successive approximations or by direct substitution. Thus, using the method of successive approximations we have obtained expressions for K to $O(c^{10})$ for the three cubic arrays. Again, in this case we found that for $\chi < 0.9$ both methods converged to the same answer. For $\chi > 0.9$, however, the direct substitution method converged rapidly whereas the successive approximations method did not converge at all. Of particular interest are the values of K when the particles are touching each other, i.e. when $\chi = 1$. For simple and body-centered cubic arrays the respective values of K are 42.1 and 170. Except for the case of nearly touching face-centered cubic arrays, all of our results agree well with those computed recently by Zick and Homsy [12].

ACKNOWLEDGEMENT

This work was supported in part by a grant by the National Science Foundation ENG-78-09241.

REFERENCES

1. Batchelor, G.K.; O'Brien, R.W.: Thermal or electrical conduction through a granular material. Proc. Roy. Soc. Lond. A355, 313-333 (1977).
2. Hasimoto, H.: On the periodic fundamental solutions of the Stokes equations and their application to viscous flow past a cubic array of spheres. J. Fluid Mech. 5, 317-328 (1959).
3. Jeffrey, D.J.: Conduction through a random suspension of spheres. Proc. Roy. Soc. Lond. A335, 355-367 (1973).
4. Levine, H.: The effective conductivity of a regular composite medium. J. Inst. Maths Applics. 2, 12-28 (1966)
5. McKenzie, D.R.; McPhedran, R.C.; Derrick, G.H.: The conductivity of lattices of spheres II. The body centered and faced centered lattices. Proc. Roy. Soc. Lond. A362, 211-232 (1978).
6. McPhedran, R.C.; McKenzie, D.R.: The conductivity of lattices of spheres I. The simple cubic lattice. Proc. Roy. Soc. Lond. A359, 45-62 (1978).
7. Meredith, R.E.; Tobias, C.W.: Resistance to potential flow through a cubical array of spheres. J. Appl. Phys. 31, 1270-1273 (1960).
8. O'Brien, R.W.: A method for the calculation of the effective transport properties of suspensions of interacting particles. J. Fluid Mech. 91, 17-39 (1979).
9. Runge, I.: On the electrical conductivity of metallic aggregates. Z. Tech. Physik. 6, 61-68 (1925).

REFERENCES (continued)

10. Sangani, A.S.; Acrivos, A.: The effective conductivity of a periodic array of spheres. To be submitted (1981).
11. Sangani, A.S.; Acrivos, A.: Slow flow through a periodic array of spheres. To be submitted (1981).
12. Zick. A.A.; Homsy, G.M.: Stokes flow through periodic arrays of spheres. (to appear) J. Fluid Mech. (1981).
13. Zuzovski, M.; Brenner, H.: Effective conductivities of composite materials composed of cubic arrangements of spherical particles embedded in an isotropic matrix. J. Appl. Math. Phys. (ZAMP) 28, 979-992 (1977).

DIELECTRIC AND ACOUSTIC RESPONSE OF ROCKS

P. N. Sen
Schlumberger-Doll Research
Ridgefield, CT 06877

ABSTRACT

A self-similar model is developed to explain the conductivity, dielectric, and acoustic response of rocks. The model explicitly takes into account the correlations that lead to a zero percolation threshold, and gives the d.c. conductivity of rock $\sigma(0) = \sigma_w(0)\phi^m$, with $\sigma_w(0)$ the d.c. conductivity of the brine, ϕ the porosity. The exponent m depends on the grain shape. Using well-known results, it is shown that a small concentration, η, of plate-like grains of aspect ratio $\delta, \delta << 1$, can give rise to a large ($\sim 10^4$) low-frequency dielectric constant ϵ_s, and explain its salinity and frequency dependences. The case $\delta < \eta$, $\eta \to 0$, corresponds to the well-known Maxwell-Wagner effect where a divergent ϵ_s is accompanied with a conductivity threshold $\sigma(0) \to 0$. The case $\delta \to 0, \eta \to 0, \delta > \eta$ gives rise to a new result, previously overlooked, with divergent ϵ_s and non-zero $\sigma(0)$. The velocity v of Biot second wave at high frequency and the fourth sound wave in ^{4}He saturated plug are given by $v = v_F/n$, v_F being the velocity in the free fluid, $n^2 = \phi \sigma_w(0)/\sigma(0)$ (exact) $= \phi^{1-m}$ (self-similar model).

Introduction

A class of inhomogeneous materials - sedimentary rocks - show a variety of interesting behavior. In this paper, a short description of a few of the properties of sedimentary rocks and explanations of those properties in terms of geometrical models are given. Most of the results presented here either have already been published, or are in the process of being published elsewhere.[1-4]

The real part of the dielectric constant, $\epsilon(\omega)$, of brine saturated rocks at low frequencies (kilohertz range) can be as great as 10^4. These gigantic values[5-6] are remarkable considering that ϵ of the individual constituents do not exceed 80: The ϵ of dry rock is about 10 and of brine about 80. Although ϵ of brine and rock separately are frequency independent up to 1 GHz, the brine saturated rock shows a great deal of dispersion. The d.c. conductivity of brine saturated sedimentary rocks obey a simple power law - known as Archie's law

$$\sigma(0) = \sigma_w(0)\phi^m \quad (1)$$

In Eq. (1), ϕ is the porosity and m depends on the grain aspect ratio, and $\sigma_w(0)$ and $\sigma(0)$ are the d.c. conductivities of brine and the rock. A first principles derivation of Archie's law has been given by Sen, Scala and Cohen[1] in terms of a self-similar model. A summary of the model is given below. The exponent m in Eq. (1) depends on grain shape. The value m = 3/2 for spheres agrees extremely well with experiments[1] on brine saturated fused glass beads.

We then show that the presence of high aspect ratio (thin plate-like) particles can explain[2-3] the experimentally observed frequency and salinity dependences and also the great values of dielectric constant at low frequencies. The shape of the particle is extremely important in general. This shows up in another context: Far infrared absorption by ultrafine metal particles is generally found much greater than that predicted for spheres. Needle-like structures formed by these spheres can explain these large absorptions. Surface effects become important at frequencies below the MHz range and also contribute to these large values of dielectric constant.

Several "potential" problems for porous media - dielectric response, thermal conductivity, etc. - are closely related. The conductivity can be used to predict the velocity of a sound mode in fluid saturated porous media.[4] This mode is known as the fourth sound in porous plugs saturated with ^{4}He and Biot slow wave in fluid saturated rocks.[7]

Surface Effects

The electrical properties of a rock-water mixture depends on the bulk electrical properties of the constituents, and then on...

   a. Electrochemical/Interfacial Effects

   b. Geometrical arrangement of the constituents

The interfacial effect is particularly prominent in clayey samples. The counter-ions on the clay surface become mobile in the presence of water and increase σ_w. These counter ions are polarizable, and contribute to $\epsilon(\omega)$. In this paper we denote the complex dielectric constant by $\epsilon = \epsilon + 4\pi\sigma/i\omega$. In the model of Schwarz,[8] which explains these large values, the basic mechanism is as follows: When an external field is applied, the interfacial charges move tangentially across the solid/liquid interface creating an excess of charge on one side, and a deficiency on the other. Since the distance, over which charges move, is macroscopic, the dipole moment induced (charge x distance) is great, hence contributing a large value of dielectric constant ($\epsilon \sim 10^3$ below 10^4 Hz). It can be shown[9] that the Schwarz model is correct only if the double layer is impenetrable to charges moving normally to the surface. Whether this is true remains to be shown. The hindered rotation of the water molecules adsorbed in the interfacial zone gives rise to a dielectric response in the MHz range, according to Schwan.[10] This is a mechanism not covered by the Schwarz model. Thus there are at least three types of contribution from the interfacial ions...

   i. Free-ion-like contribution which gives the finite d.c. conductivity

   ii. Polarization of the counter-ion layer

   iii. Polarization of hindered water molecules

The detailed understanding of the interfacial effects needs more theoretical and experimental work, but the geometrical effects are better understood. The rest of the paper deals with geometrical effects.

Geometric Effects

In general, the macroscopic properties of a composite material containing phases with very different physical properties, depend not only on the volume fractions of the constituents, but are extremely sensitive to the geometry and topology of the boundary surfaces between the phases. The parameters of typical boundary surfaces in a mixture are not related, in any simple way, to the lowest order point distribution functions for the constituents, so that the mathematical problem of calculating the overall properties of the system is very ill−posed.[11]

The importance of the geometrical effects can be simply illustrated by the following *exact* results. The dielectric constant of a layered medium (with thicknesses smaller than the wavelength) or a medium consisting of tubes, when a field is appied parallel to the interface, is given by,

$$\epsilon^* = \sum_i f_i \epsilon_i^* \qquad (2)$$

where f_i denotes the volume fraction of the i-th phase of dielectric constant ϵ_i^*. But when the field is applied perpendicular to the interface of the layered medium, ϵ^* is given by

$$\epsilon^* = \left[\sum_i f_i / \epsilon_i^* \right]^{-1} \qquad (3)$$

If the layers are made up to conducting and insulating materials, in the first case we obtain a non-zero $\sigma(0)$, whereas $\sigma(0)$ is zero for the latter. Similarly, the dielectric constants are drastically different in two cases. The importance of the interface-shape can be further illustrated. The potential ϕ at a point $\mathbf{r}$, is given by[12]

$$\phi(\mathbf{r}) = \phi_0(\mathbf{r}) + \int \mathbf{P} \cdot \nabla' \frac{1}{|\mathbf{r} - \mathbf{r}'|} d^3 r' \qquad (4)$$

The term ϕ_0 is from the fixed external sources (e.g., the charges on condenser plates) and $\mathbf{P}$ is the dipole moment per unit volume. Integrating by parts, (4) gives

$$\phi(\mathbf{r}) = \phi_0(\mathbf{r}) - \int \frac{\nabla' \cdot \mathbf{P}(\mathbf{r}')}{|\mathbf{r} - \mathbf{r}'|} d^3 r' + \sum_\alpha \frac{\mathbf{P} \cdot d\mathbf{s}_\alpha'}{|\mathbf{r} - \mathbf{r}'|} \qquad (5)$$

The sum over α denotes a sum over all the interfaces. Here we envision the material to be composed of non-overlapping cells, each cell having a constant polarizability. If $\mathbf{P}$ is uniform in each cell, which is the case in the single site approximations (both coherent potential approximation or average t-matrix approximation discussed below), *only the* interfacial terms contribute. In that case of $\nabla \cdot \mathbf{P} = 0$ in each cell, the field is given from (5) to be

$$E_i = -\partial\phi/\partial x_i = E_{oi} - 4\pi \sum_\alpha L_{\alpha ij} P_{\alpha j} \tag{6}$$

and $L_{\alpha ij}(r)$ is the generalized depolarization factor of the αth grain

$$L_{\alpha ij} = -\frac{1}{4\pi} \int \frac{r_i - r_i'}{|r - r'|^3} ds'_{\alpha j} \tag{7}$$

which depends on the grain shape. Thus, the shape of the interface is of paramount importance in determining the dielectric constant of an inhomogeneous material.

Equation (4) forms the basis of systematic iteration schemes for evaluating the dielectric constant of the composite material.[14] Since P is related to E via a constitutive relation $P = (\epsilon^* - 1)E/4\pi = \alpha^* E$, (4) gives an integral equation for E. This also is the starting equation for derivation of the Clausius-Mossotti formula following Ewald and Oseen.[15] The field at a given point is given by the external field plus the sum of fields due to each cell. In order to compute the field produced by each cell, we need to obtain the field which the cell itself is subjected to. In a single site approximation, each cell is assumed to be surrounded by an average medium of dielectric constant ϵ_o^*. In the average t-matrix approximation (ATA), this is taken to be the host material. The effective dielectric constant ϵ^* is computed by requiring that this material of dielectric constant ϵ^*, when embedded in ϵ_o^*, will give the same scattered field on the average. The other common single site approximation, coherent potential (CPA) or effective medium approximation (EMA) obtained by requiring that ϵ^* is such that the scattering vanishes on the average.

The formal iteration of (4) has been developed by many authors. These techniques have been summarized by M. Hori.[14] I give here a simple derivation due to Maxwell.[16] This derivation clarifies all the points which form the basis of the formal perturbation analysis and truncation schemes mentioned above. Suppose, spheres of radii a_i having dielectric constant ϵ_i^* are embedded in an assumed host ϵ_o^*. The total potential due to these spheres subjected to a field $E_o \hat{z}$ at infinity is

$$\phi = \sum_i \frac{\epsilon_i^* - \epsilon_o^*}{\epsilon_i^* + 2\epsilon_o^*} a_i^3 E_o \frac{r\cdot\hat{z}}{r^3} - E_o(r\cdot\hat{z}) \tag{8}$$

this is equated to the potential due a large sphere of dielectric constant ϵ^* and radius R that encloses all these small spheres:

$$\phi = \frac{\epsilon^* - \epsilon_o^*}{\epsilon^* + 2\epsilon_o^*} R^3 E_o \frac{r\cdot\hat{z}}{r^3} - E_o(r\cdot\hat{z}) \tag{9}$$

In (9), ϵ^* is the effective dielectric constant of the mixture. Equations (8) and (9) give the ATA results

$$\frac{\epsilon^* - \epsilon_o^*}{\epsilon^* + 2\epsilon_o^*} = \sum_i f_i \frac{\epsilon_i^* - \epsilon_o^*}{\epsilon_i^* + 2\epsilon_o^*} \tag{10}$$

Here f_i is the volume function of the i-th phase a_i^3 / R^3. If we require self-consistency i.e., $\epsilon_o^* = \epsilon^*$, then (8) gives, instead of (10)

$$\sum_i f_i \frac{\epsilon_i^* - \epsilon^*}{\epsilon_i^* + 2\epsilon^*} = 0 \qquad (11)$$

In ATA, (10), ϵ_o^* is generally taken as the majority substance. Equation (11) implies, via Eq. (8), that the total potential produced by the spheres add up to zero. The key point in the derivation above is that the local field seen by each sphere is the same and is not affected by the *proximity* of the other spheres. This is the basic deficiency of the single-site theory. We describe below the self-similar model[1] which goes beyond the dilute limit. There are, of course, competing theories but we do not focus on them in this article.

The First Principles Derivation of Archie's Law: Self-Similar Model

In reference 1 we developed a theory for dielectric response of water-saturated rocks based on a realistic model of the pore space. The absence of a percolation threshold manifest in Archie's law, porecasts, electron-micrographs, and general theories of formation of detrital sedimentary rocks indicate that the pore spaces within such rocks remain interconnected to very low values of the porosity ϕ. In the simplest geometric model for which the conducting paths remain interconnected, each grain is envisioned to be coated with water. The dielectric constant of the assembly of water-coated grains is obtained by a self-consistent effective medium theory.

Using equation (11), the self-consistent dielectric constant of the rock made up of coated spheres is given by

$$0 = \sum_i f_i \frac{\epsilon^* - \epsilon_{cs}^{(i)}}{2\epsilon^* + \epsilon_{cs}^{(i)}} \qquad (12)$$

Here f_i is the volume fraction of the i-th coated sphere in the assembly, i.e., volume of the i-th sphere divided by the total volume of the rock. The dielectric constant of a sphere i of material ϵ_m^* coated with water ϵ_w^* is[17]

$$\epsilon_{cs}^{(i)} = \epsilon_w^* \frac{\epsilon_m^* + 2\epsilon_w^* + 2\eta_i(\epsilon_m^* - \epsilon_w^*)}{\epsilon_m^* + 2\epsilon_w^* - \eta_i(\epsilon_m^* - \epsilon_w^*)} \qquad (13)$$

Here η_i is the volume fraction of the inner sphere. In general, f_i and η_i are unrelated. The simplest case arises if we assume that $\eta_i = 1 - \phi$ is the same for all the spheres. Then separating real and imaginary parts of (12) gives in the d.c. limit

$$\sigma(0) = \sigma_w(0) \frac{2\phi}{3-\phi} \qquad (14)$$

Equation (14) gives $\sigma \propto \phi^m$ with $m = 1$ for small ϕ. The empirical evidence for an exponent m greater than unity is overwhelming, both from laboratory and field data. Accordingly, we must regard the model of a sphere coated with a spherical shell of water of fixed proportions as overly simple.

The single-site or mean-field approximation such as CPA fails to take into account a variety of local environments of each type of rock grain. The systematic way around this problem is to consider bigger and bigger clusters, where a grain is not only surrounded by water but also surrounded by other rock grains. In practice this problem is extremely difficult and has been solved numerically only for a few special cases. We proceed here by a very simple intuitive method of incorporating the clustering effects in a single-site effective medium theory. This comes from the observation that each grain is not only surrounded by water but also by other grains which in turn are surrounded by others in a nested manner. This led to a self-similar model where the continuity of the conducting phase is guaranteed.

First we add a few grains of rock of any size to water. As long as the sizes are smaller than the smallest wavelength involved, the sizes are irrelevant. Then use this mixture to coat some new grains, and so on. At each step we add a small amount of grain, and we determine the dielectric constant self-consistently by generalizations of eqs. (13) and (12). This gives for an infinitesimal increase of grain volume,

$$\frac{d\epsilon^*}{\epsilon^*} = -\frac{d\phi}{\phi}\frac{\epsilon_m^*-\epsilon^*}{\epsilon_m^*+2\epsilon^*} \tag{15}$$

Integrating (15) with the boundary condition that $\epsilon^* = \epsilon^*_w$ at $\phi = 1$ gives

$$\left[\frac{\epsilon_m^*-\epsilon^*}{\epsilon_m^*-\epsilon_w^*}\right]\left[\frac{\epsilon_w^*}{\epsilon^*}\right]^{1/3} = \phi \tag{16}$$

The limiting case (as the frequency, ω, tends to zero) of equation (16) gives Archie's law. Separating the imaginary part gives

$$\sigma(0) = \sigma_w(0)\phi^{3/2} \tag{17}$$

For anisotropic rock with grains of fixed shape and orientation, equation (17) is easily generalized. For ellipsoidal grains with a depolarization ratio L (L is a function of grain shape[1-4,12,13] we find, under fairly general conditions,[1,3,4,18,19]

$$\left[\frac{\epsilon_m^*-\epsilon^*}{\epsilon_m^*-\epsilon_w^*}\right]\left[\frac{\epsilon_w^*}{\epsilon^*}\right]^L = \phi \tag{18}$$

This, in the d.c. limit, gives

$$\sigma(0) = \sigma_w(0)\phi^{\frac{1}{1-L}} \tag{19}$$

For isotropic rock, in which the spheroidal grains are randomly oriented, the cementation exponent is given by

$$m = \int \frac{5-3L_s}{3(1-L_s^2)} P(L_s) dL_s \qquad (20)$$

Here $P(L_S)$ is the probability distribution for the depolarization factor L_S along the symmetry axis. Equation (16) is identical to that obtained by Bruggeman.[20] Bruggeman's procedure was generalized by Veinberg[18] for randomly oriented spheroids (and was applied to rocks by Mendelson and Cohen[19]) to obtain (18) and (20).

The procedure used by Bruggeman[20] and Veinberg[18] did not employ the coated spheres technique. They introduce grains, by an infinitesimal amount at each step, into the previous mixture and use the effective medium approximation (EMA) at each step. Since only an infintesimal amount of grain is introduced at each step (far below the percolation threshold), the medium remains conducting at each step. Furthermore, for coated spheres, EMA is equivalent to uncoated spheres with the Clausius-Mossotti (ATA). But, ATA in turn is the same as EMA in the dilute concentration limit. Thus the procedures of Sen, Scala and Cohen[1] and of Bruggeman[20]-Veinberg[18] are similar provided that we use the correct boundary condition, e.g., start with water at step zero.

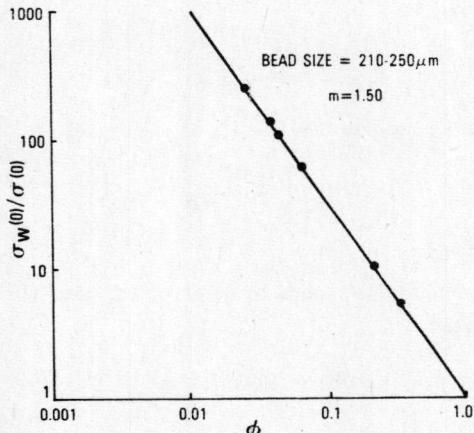

Fig. 1 - D.C. Conductivity of Glass Beads; Dots: Expt; Line: Theory (17)

The conductivity of the fused glass beads at low frequencies agree very well with equation (17) with an exponent of $m = 1.5$ (Fig. 1). Theory also predicts $\mathrm{Re}\,\epsilon^*$ for many cases extremely well.[1]

Frequency/Salinity Dependence and Large Values of Dielectric Constants: Thin Plate Effects

Despite the successes of the models discussed above, we have failed to predict the most spectacular property - the dielectric constant ϵ reaches values as high as 10^4 at low frequencies. In